MICROBIOME MODIFICATION

MICROBIOME MODIFICATION

The Superorganism for Good Health

Patricia Worby, Ph.D.

MERCURY LEARNING AND INFORMATION

Boston, Massachusetts

Publisher: David Pallai
MERCURY LEARNING AND INFORMATION
121 High Street, 3rd Floor
Boston, MA 02110
info@merclearning.com
www.merclearning.com
800-232-0223

P. Worby. *Microbiome Modification: The Superorganism for Good Health.*
ISBN: 978-1-50152-244-4

Library of Congress Control Number: 2024930832

242526321 This book is printed on acid-free paper in the United States of America.

Our titles are available for adoption, license, or bulk purchase by institutions, corporations, etc. For additional information, please contact the Customer Service Dept. at 800-232-0223(toll free).

All of our titles are available in digital format at academiccourseware.com and other digital vendors. The sole obligation of MERCURY LEARNING AND INFORMATION to the purchaser is to replace the files, based on defective materials or faulty workmanship, but not based on the operation or functionality of the product.

CONTENTS

PREFACE

This book originally came into being due to an overwhelming urge of mine to contribute to the new understanding of the microbiome that was unfolding just before the coronavirus pandemic. I could not have foreseen the effect that extraordinary event would have on the population, but it helped to highlight the importance of the virome (part of the microbiome) to many more people. Unfortunately, it also contributed to a negative connotation that all microbes are dangerous – an outlook this book very much wants to argue against. The fact that the health of your inner garden (your gut) determines your susceptibility to disease was ignored in the pandemic response which was solely focused on a war against one virus.

As I have watched this play out, I have become even more determined to show people the magic and wonder of our inner microbial community and how they contribute to health. I wanted to do so in a way that everyone could access without needing familiarity with biological science. Hence, I have included a glossary of all scientific terms and some of the concepts are repeated in novel ways throughout, as it is my belief that the more times you hear something the more likely you are to remember it. But overall, my focus is on what *you* can do to heal your microbial community without relying on an overstretched and sometimes inadequate health service.

In the first chapter, I outline the distribution and diversity and some essential concepts including gut permeability, the microbial contribution to immunity, digestion and absorption, the gut-brain, nutrient synthesis, and genetic regulation.

Chapter 2 outlines their fascinating role in our evolutionary history to complex animals via mitochondria – our energy and cell signalling organelles that use quantum processes to alter energy output and change the expression of genes to control metabolism epigenetically. I also consider the quantum mechanics of the cell microtubule structure which is rarely taught in biology. Chapter 3 details the threats to health from both outside (agricultural practice, environmental toxins, pharmaceuticals, and how microbes protect themselves from attack. I also cover the less well-known subjects of parasites and how stress changes the microbiome to an unhealthy one.

Chapter 4 covers good nutrition as the foundation of health, looking into subjects such as good fats, probiotics and fermented foods, and dietary regimes that help regulate (gluten free, ketogenic diets, and intermittent fasting). We also consider how home juicing and fermenting can alter your microbiome for the good and finish the chapter with a consideration of phytonutrients and how they work at a cellular level to harvest light and boost your energy and detoxification systems.

Chapter 5 considers lifestyle factors including the importance of clean water, sunlight, exercise, good sleep, and stress reduction in a healthy body. We look at medicinal herbs and essential oils to support detoxification and end with the special considerations of specific life stages life conception, birth, and child development. I introduce the emerging field of psychobiotics in mental health.

Chapter 6 focuses on the diseases of an imbalanced microbiome (allergies, auto-immune disease, cancer, chronic fatigue/ long covid, IBS, celiac disease, and more surprisingly "mental" diseases such as Alzheimer's, anxiety, and depression) and how one may treat them by changing your microbial balance.

In the last chapter, I explore the bigger picture of health and wellbeing – how vested interests keep us stuck in a disease model of health based on germ theory vs. the more hopeful terrain model (for example vaccines vs. building natural immunity – the naturopathic approach). I finish with how mind and matter collaborate and how by healing ourselves psychologically/emotionally we build a more resilient microbiome. This truly is 21st century healthcare – personalized and proactive and a far cry from the model we are currently adopting. It is my hope that you will find yourself both inspired and motivated to take back your power and create health and wellbeing for you and your family.

ACKNOWLEDGMENTS

I would like to acknowledge my partner, Jill and our furry friend, Sidney for the support and occasional distraction. Special thanks to my project manager, Jennifer Blaney and in particular, James Walsh at Mercury Learning for hand-holding me through the process.

Patricia Worby
January 2024

CHAPTER 1

THE IMPORTANCE OF THE MICROBIOME

Microbiome Definition and Variation

The microbial cells present in our gut and other parts of our body, acting as a living shield to help protect us from opportunistic pathogens and keep us healthy, are collectively termed the *microbiome.* The fascinating truth is you are a host for trillions of bacteria and other essential microorganisms! They number at least ten times more than the cells of your body and, perhaps more importantly, contribute between 150–500 times the genetic information.[1]

There is a phrase from Egyptian philosophy: "as above, so below" that refers to the macrocosmos mirroring the microcosmos.[2] I have pondered the meaning of this mystic belief for many years, but it intrigued me again recently when considering the microbiome. Indeed, given our recent discovery that the microbiome (genetic information) outnumbers your own cells by a vast number (some say up to 500:1), it would seem to be a good analogy with this philosophy; the outside being inside, as it were.

Why is it important that they outnumber us so? It seems it may have conferred evolutionary advantage to us. It's a bit like the concept of Russian dolls: in addition to the trillions of microbes (mostly bacteria) within our

---

[1] There are 10^{14} cells in your body and 1016 microbial species that live on and within us. That's 10,000,000,000,000,000 roughly!

[2] The Emerald Tablet, part of the Ancient Egyptian Hermetica, considers this to mean God is within. But we would broaden it to mean "life dwells within us."

bodies, each with their own DNA, there are even smaller microorganisms that live within the bacteria, called *bacteriophages* which themselves outnumber the bacteria 10:1. Each bacteriophage carries genes within its cell structure, so you can see there are many hundreds of times the genetic component of your own cells. Thus, 99% of your genetic information (the *genome*) is in the gut and other organs populated by these microbes—called, collectively, the microbiome (short for "microbial genome").

Despite the fact we only ever consider them as present in the gut, they are everywhere in the human body (except, perhaps, the brain). Furthermore, their distribution and constitution vary between regions of the body. This is deeply significant as we will see.

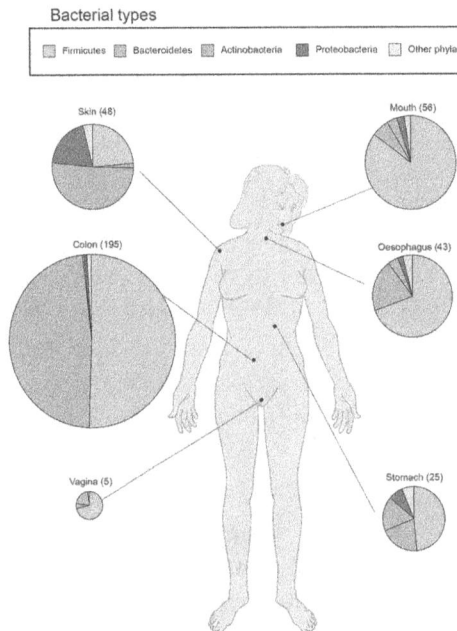

Bacterial types

| Firmicutes | Bacteroidetes | Actinobacteria | Proteobacteria | Other phyla |

Skin (48)

Mouth (56)

Colon (195)

Oesophagus (43)

Vagina (5)

Stomach (25)

Figure 1.1. The variation of species in the microbiome.

They may hold the key to many diseases, including neurodegenerative illness and other conditions, because what genes get expressed depends on the balance of the microorganisms you have living within you (the host). Moreover, this is as individual as you are. When we sample blood or urine, we are actually sampling the *collective* metabolism of the host-microbiome system. This is as big a revolution in understanding as that of quantum physics was for the physical sciences at the turn of the last century. Thus, we could say it is a quantum revolution in biology.

Using this new understanding of the gut as a *postnatal organ system*,[3] consisting of a dynamic ecosystem of microorganisms, we see that each contributes an essential component. The microbiome weighs an impressive 4lbs.–5lbs. (1.5kg. or about 2% of our body weight) in mass and so is worthy of consideration as is any other organ of the body. But, beyond biology, it also is a living record of our evolutionary history as, according to respected US doctor, Dr. Raphael Kellman, it represents "a primordial remnant of the past; it carries the essence of what makes us what we are." Because of its complex composition of rapidly mutating organisms,[4] it is constantly swapping genetic information with us and each other. This allows it to respond to changes in the environment (both internal and external) and means it is in constant communication with the internal milieu. Naturopaths call this the *bioterrain* and it is useful to think about this as the "playing field" within which our body's cells and the microbiome organize themselves. Altering this terrain changes the type of organisms that dwell there.

Hence, in their complex interplay of species and genetic transfer, they become "the repository of all knowledge about what it takes to make a human" [Sayer16]. From this perspective, we can consider that microbes carry within us the essence of life. Their *metabolites* (products of their biochemistry) are the footprint of that inner ecology. When we do a blood test, for instance, we measure not only those metabolites that we produce within our *own* cells but also those made by the bacteria within. So, our particular microbial metabolic output gives us a lot of information about our current state of health, but it could also be considered like an excavation into the past. In other words, it is a living collection of your interaction with the microbes passed down to you from your mother[5] and other family members, not to mention your environment generally.

Most metabolites come from our gut bacteria not from us. This is a surprise to most people. Indeed, when I did my biochemistry degree in the 1980s, this was never mentioned. We studied complex biochemical pathways of the production of proteins and neurotransmitters, hormones, and energy via such delights as the Kreb's cycle, which we students had to learn by rote. But the contribution of the gut flora was not discussed at all; we assume it was not known. With the knowledge that we now have, we perceive another avenue of healing that is available to us: by identifying

---

[3.] It develops after birth rather than before.

[4.] Bacteria can reproduce much faster than the cells in your body as the structure of chromosomes is much simpler.

[5.] The microbes are inoculated from the birth canal during birth.

the unique make-up of our gut flora by microbial testing, we can design our personal prescription of *probiotics* (introduced microbes within the body) to provide those elements that are missing or out of balance. This constitutes a form of personalized medicine that has hitherto been only a dream. However, there is an important proviso here: probiotics are not like antibiotics. You can't just use them randomly in the mistaken belief that they "must do some good." You must address the issues that are causing the imbalance at the same time.

Leaky gut is one factor that seems to occur commonly in western countries where we get most of our energy from carbohydrates[6] (as fat consumption has reduced). We must heal the gut first by healing this leakiness—reducing inflammatory foods, adding nutrients, and keeping hydrated. Also, we must address the hormonal factors and heal the thyroid (typically either low amounts of thyroid hormone or a hypothyroid condition where conversion to active T3 hormone is not adequate). This is covered later in the book, but we must be clear, healing the gut and rebalancing the flora are not the same. You can't just ingest probiotics as a panacea for all issues. We need to be more subtle than that.

Perhaps, though, by looking at our bodies as a community rather than an "I/me," we can gain a new understanding. Indeed, US naturopath Raphael Kellman believes that viewpoint leads to a new paradigm in medicine that encourages more "sensitivity to the needs of the other."

Another influential expert, David Perlmutter, agrees that "every aspect of our health and physiologic function is dependent on the health of the organisms that live within and on us." We are a triumph of symbiotic living and the conclusion many are rapidly coming to is that "to be human is to be multispecies" [Hutter15, para. 2]. If we expand the horizon outside the body, we can consider the wider external microbiome too; for example, the atmospheric, ocean, and soil microbiome. We are destroying that as well, through lack of care and respect for its intrinsic importance to our survival. Indeed, our disrespect for our outer world challenges the repair mechanisms of our environment. At its most obvious, the results of this disaffection are affecting our climate in ways that could threaten our future existence. The same is true within the body as we destroy the terrain that bacteria have evolved within for eons. As you can now see, an understanding of the microbiome has implications way beyond the body. As above, so below. As within, so without.

[6.] Which are considered proinflammatory when combined with poor gut flora.

The Bioterrain and Nutrition

Let's return to the environment of your body and consider the bioterrain (environment). This constitutes everything in your body, inside and out: blood, skin, gut, and so on. The quality of this environment has a lot to do with the way that you live and eat. Modern agricultural methods have prioritized yield above nutrient density to the detriment of our planet and ourselves. Interestingly, we are the first generation to live potentially shorter lives than those of our parents. You will hear the idea that there is no reason we could not all live to 100 trumpeted in the media. This is true theoretically, but the reality is we may live longer lives now than the early years of the twentieth century, when infectious diseases were the main killer, but our years of *healthy living* are decreasing all the time. Increasingly, we spend our latter years with chronic illness of one sort or another, dependent on a multitude of drugs to keep us alive; this is hardly living, more a slow, painfully declining existence.

The main cause of the epidemic of chronic disease is malnutrition. By that we do not mean lack of (protein) food as is commonly understood in developing nations. We refer instead to the dearth of nutrients in modern processed foods. We are starving for nutrients, but they are not present in most modern foodstuffs; the source foods from which they are made contain few and then processing destroys the remainder. Grains are a good example. Grown on intensively farmed soils, they become depleted and weak. They need more and more artificial fertilizers to encourage growth and a barrage of pesticides to prevent infestations. While this improves yield it does not improve quality. It becomes a downward cycle.

So, our first priority when addressing the problem of a sick bioterrain, is to increase the nutrient density of our foods. Most of us can't take on the agribusiness industry, but we can, thankfully, still obtain nutrient-dense foods grown on organically farmed land. If we don't have the money to invest in relatively expensive organic food, we can grow our own rather cheaply. We can also cook from scratch using foods with only one ingredient. Refer to an integral book on the subject titled *In Defence of Food* by Michael Pollan [Pollan09]. It discusses this in more detail. Unfortunately, this means most of the shelves at the supermarket become "out of bounds." There aren't many nutrients in packet cereals, sweetened yogurts, ready-made meals, fizzy drinks/ sodas, and confectionery, despite what marketing tells you. In fact, if anything has to be "fortified" that's a signal it is junk food to start with.

Thus, the first course of action in any healing program will be to change the foods you eat by selecting foods with natural high-nutrient densities and known provenance. Organic and/or locally grown foods offer the best option for this. Long-distance travel and storage of conventionally grown foods means most of the vitamins and minerals are lost even before they arrive at your home, let alone those lost in cooking. You need to start with organic food, rather than resorting to supplements on top of a poor diet. There is a reason the manufacturers of vitamins say, "not a substitute for a good diet!" This is especially true of supplements as a solution for poor digestion (or indeed the other less naturopathic remedies like antacids and other over the counter medicines). As nutritionist, Emma Chapman-Sharp, has commented "we need to move away from the era of anti-everything"; antibiotics, antacids, antidepressants, and so on are simply a "sticking plaster" over the problem that does nothing to address the root cause.

This truth is even more relevant for probiotics. Simply adding extra bacteria (which is what probiotics are) in the form of a pill does nothing to change your gut flora permanently if you don't create the right *terrain*; they will simply not stick around. Being a dynamic community, they need to have the right *environment* to proliferate and survive. Otherwise, they are transient and only last while you are taking the probiotic (a reason people feel better on them usually, but any benefit disappears when they stop). I'm not saying that probiotics are not useful, in fact they are a staple of any healing program, but it has to be alongside significant changes in your lifestyle and eating habits. This is a community we are talking about here. Very much like human communities—if you don't keep their needs met, you get a host of undesirables moving in.

Another important task in the healing program is to increase your digestive capabilities. Over time on a western depleted diet, (sometimes called the standard American diet [SAD]), your digestive function begins to reduce. Sadly, one of the highest energy requiring functions of the body is digestion itself and it needs good quality nutrients to work properly. Therefore, you can see that once you start to lose digestive power, your ability to assimilate your food becomes less and then you enter a vicious cycle of lowering function. Two essentials of good digestion are *digestive enzymes* and *hydrochloric acid* of the stomach. Let's look at the process in more detail.

Eating and Stress

In order to make energy from our food,[7] we need to ingest and break down the raw ingredients: protein, carbohydrates, fat, plus a good number of vitamins and minerals. To do this, we have a complex set of enzymes (proteins that help to break down food), hormones (that stimulate and regulate appetite for instance) and the right cofactors (minerals like selenium, magnesium, B vitamins, and so on) to run the chemical reactions where different molecules are transformed. You are not what you eat, but what you *assimilate*; you have to be able to break the foods down properly to bring them into the body and reuse them. Remember, we have to create our tissues from the raw materials of food.

Let's look at the first part of digestion—chewing. Now it's true to say, most people do not pay enough attention to this important aspect of digestion. Modern life does not encourage us to find time to sit down and take time to eat. We are constantly on the run, fitting in eating while doing other things (multitasking and snacking). This is disastrous for our digestive capacity. When we are not paying attention to eating, we tend to do it fast, we don't chew but gulp, interfering with the mixing with saliva (which is essential for beginning digestion in the mouth in addition to alkalizing the food). We don't breathe properly either; so, it becomes shallow and fast rather than slow and deep from the belly.

Most of us now don't take our time when cooking either; the average preparation time has dropped to twenty minutes for most meals from forty minutes in the 1950s [Zick10]. Neither do we tend to eat together with others; in families today people often eat separately, in front of screens. It's no wonder we are not paying attention to what, and how much, we eat! Eating socially has been shown to slow down the process of eating as we talk between courses. This means we are more likely not to overeat as we take time to assess how full we are. But in any case, *eating in a state of relaxation and pleasure is vital for proper digestion.* It is one of the factors that explain the "French Paradox"; the fact that despite a diet of high saturated fat the French have lower levels of heart disease [Ferrières04]. Eating with others is much more conducive to good digestion (as is consumption of moderate amounts of red wine—with food).

[7.] Making energy from food not sunlight is what defines an animal as opposed to a plant, but this has recently been called into question as it seems animals do use light energy to produce energy; not through light hitting leaves but using the photon capture of eaten plant pigments as described in a later chapter.

Stress hugely affects how we digest; there is a social movement called "slow food" that advocates more time over preparation and eating, and the movement is gaining ground as an antidote to our 24/7 fast-food culture. In another development, mindful or "conscious eating" is promoted by such people as Dr. Gabriel Cousens.[8] He particularly believes in the preparation of food as an act of love, and we can all attest to the different quality of meals that have been cooked that way. If we are lucky enough to have grown up in a family where we were cooked for, and the cook (whether that was your mother, father, or other family member) *enjoyed* cooking—you will remember how much better it tastes. I grew up at a time when school dinners were like home cooking and, although we moaned at the time, they were cooked properly and with care. What we perhaps haven't appreciated up until now is that our microbiome responds to this style of preparation better too. The absence of stress hormones affects them and makes them respond with better balance and digestive efficiency (microbes help us digest food and keep our gut lining healthy).

So, it is important to deal with your stress and take time out for eating, both at work and at home. Stress management isn't just a luxury but an essential for healthy living. For many of us our stress levels are permanently high, but we are so used to it we don't even realize it. We are "tired but wired" due to the flooding of the body with stress hormones like cortisol. In order to improve our long-term health, we need to both reduce the stress we're under (by sitting at a table to eat for instance, not a desk or in front of a TV or screen) but also, we need to reprogram our brains in the way in which they *deal* with stress. And we are not just talking about the obvious conscious stressors of poverty, family conflict, job stress, for instance. For many people, the stresses are *unconscious*: internal conflicts, actions inconsistent with our value systems, people who trigger us based on our childhood beliefs and responses. All of these factors cause a lifelong dangerous stress cycle that will predispose us to chronic disease [WHO03]. Indeed, stress is the biggest killer in the western world although the diseases we die of will not be labeled as such; heart disease and cancer have stress at their heart. My first book, *The Scar That Won't Heal* dealt with this in more depth.

Protein Digestion

The body is incredibly clever in how it is designed to create the ideal conditions for the breakdown of different constituents of food; the mouth

---

[8.] www.treeoflifecenter.com. See also G. Cousens, *Conscious Eating* (*Essene Vision Books, 1995*) *for more information.*

is slightly alkaline to enable efficient carbohydrate digestion—largely as a result of mixing with saliva. But in order to breakdown protein a very different environment is required. After the food has been partially broken down by enzymes present in saliva (amylase), it has the first contact with the strong acid of the stomach. The stomach produces its own hydrochloric acid (HCl) from cells present in the stomach lining. But they also require certain cofactors to be present; one of these is called, rather mysteriously, "intrinsic factor" which likely was named as the original discoverer didn't know what it was composed of (it is in fact a protein itself) [Wikipedia].

For the stomach to protect itself from self-digestion, it must have some special mechanisms in place; there are three ways it does this:

- pepsin
- bicarbonate
- mucus secretion

The stomach secretes an inactive precursor called pepsinogen which activates to pepsin (the protein digesting enzyme) when in contact with the HCl of the stomach. This is then balanced with bicarbonate secretion (alkaline to the acid of the HCl) and a thick mucus secretion from the goblet cells of the stomach lining, preventing the stomach from digesting its own cells.

The other important point about the strength of the HCl in the stomach has nothing do with digestion itself is that it *helps to prevent disease*. A highly acidic stomach kills off any potentially harmful bacteria that are present in the food from surviving beyond the stomach. So, high acidity is a defense against pathogens and needs to be preserved. Unfortunately, as previously mentioned, stress and age affect the ability of the stomach to produce HCl and as acidity declines so does this protection. Those worst affected tend to be middle-aged females, due to hormonal instability and thyroid issues that are common in this age group (and may be undiagnosed until the onset of reflux or some other digestive issue). If you find that you are always the one to come down with a stomach bug while those around you are not affected, or you feel bloated or are suffering reflux after eating, a lack of HCl could be your issue. Most GPs however tend to assume that it is an overproduction of acid and may misprescribe antacids for reflux. Proton pump inhibitors (PPIs) are also commonly given even though they have many side effects including overgrowth of pathogenic bacteria in the colon and osteoporosis and heart disease if used long term.[9] Getting digestion right at the top end maximizes digestion lower down into the gut. Therefore, making sure your diet is nutrient dense and full of live foods is really key.

Raw and Live Foods

In the past when we evolved as hunter-gatherers in the open air, our food would have been harvested when it was available. It was thus local, seasonal, and fresh. The advent of cooking seems to have made more nutrients available to us quicker as we no longer had to chew so long and hard to digest our food—cooking changes the chemical composition of food so that it is broken down more easily. However, if the food is nutrient deficient and old, sprayed, or otherwise adulterated, cooking simply reduces the nutrient content even more. This is why the Paleolithic ("Paleo") and raw food movements have arisen and claim such beneficial effects.

A strict Paleo diet only allows those foods that were available to man before the advent of agriculture. Thus, it spurns grains and dairy products, concentrating on nuts, seeds, berries, meat, and fish. Many people who have switched to this kind of diet claim amazing benefits; but it is not for everyone. Some people find it too restrictive and those with a low thyroid (who therefore have difficulty heating up their body sufficiently and producing much energy) find it makes them feel worse.

Raw food diets are an alternative that concentrate on green (plant-based) foods with live enzymes (present on and in the food) to help digest it. Nothing is cooked above 114 °F (40 °C) so as to preserve the action of these enzymes (that are different from the enzymes present in your body naturally). In order for us to be able to digest these well and get them in sufficient quantities (particularly to meet our protein requirements without meat or fish), these diets rely more on blending vegetables and fruits so that we increase our consumption dramatically. There are many exponents of these raw diets, too numerous to mention here,[10] but just to note there has been a recent backlash against them saying they are trendy and unnecessary, but they contain much of value even if they don't suit everyone.

[9.] PPIs may increase the risk of Clostridium difficile infection of the colon. High doses and long-term use (one year or longer) may increase the risk of osteoporosis-related fractures of the hip, wrist, or spine. Prolonged use also reduces absorption of vitamin B12 (cyanocobalamin). Long-term use of PPIs has also been associated with low levels of magnesium (hypomagnesemia). Analysis of patients taking PPIs for long periods of time showed an increased risk of heart attacks. http://www.medicinenet.com/proton-pump_inhibitors/

[10.] See the websites list in the Resources section at the end of this volume for more details.

Germ Theory and the Limitations of Modern Medicine

We have a crisis within Western medicine—we simply can't afford it. The model upon which it is based is deeply flawed and is costing us a fortune as our overfed, but malnourished population get sicker, and the drugs used to treat the diseases of overconsumption (obesity, diabetes, heart disease, etc.) get ever more expensive. Instead of a healthier population, things are going in the opposite direction. In the United States, for instance, despite billions of dollars spent in health care, the nation continues to get fatter and sicker year after year. This is repeated in most western nations as the same issues of adulterated and nutrient-poor food, social stress, and toxicity impact our natural systems of defense and balance. Much of this goes back to the development of Western medicine which began in the early nineteenth century with the decision to base medicine on *curing the external agent of disease in separate compartments of the body*, rather than looking at whole body system imbalance. This is known as *germ theory* or the *infectious disease model* (IDM).

At this time, little was known about the specifics of disease-causing agents. Prior to the work of Louis Pasteur, it was believed "bad air" or miasmas caused disease. But, with the advent of the first microscopes, it was possible for scientists like Pasteur to view tiny organisms isolated from diseased bodies as the cause of disease. From his theories, we have the ideas of vaccines, and most of modern medicine is based on this model of "one organism/one disease," even though Pasteur later recanted this theory.[11] But his was not the only theory, it just happened to be the most popular. Another French "gentleman scientist," Claude Bernard, argued with Pasteur and instead claimed "that it's all about the terrain." Even Pasteur eventually changed his mind and on his deathbed agreed "the terrain is everything" [Appleton02].

This terrain theory posits that disease is created by imbalance first and foremost. How the bacteria are integrated *as a whole* is more important than isolating bugs and treating them—except in instances of acute infection—and even then, we have misunderstood some of the science. Most chronic (long-lasting) disease comes about from a "bad infrastructure" or balance within the bioterrain which means the body cannot fight the pathogenic bacteria. We are not so much invaded from without but from within.

[11.] On his deathbed Pasteur is said to have claimed "the microbes are nothing; the terrain is everything."

The infectious disease model is not only seriously flawed, but it is extremely dangerous. We have been encouraged to strike microbes out of existence, both in medicine, the public realm and in the home. The proliferation of antibacterial cleaners and hand-washing gels means we kill the good bacteria that we need for health! Avoid them at all costs and instead use natural antibacterial compounds like lemon juice and those naturally found in wooden chopping boards.[12] Good kitchen hygiene is essential, certainly. You need to separate the preparation of dead cadaver foods (meat and fish) from vegetables. Even frozen foods contain bacteria and so defrosting in separate containers is essential. Ready meals, commonly sold in the UK and US, are often a source of infection [Willacy23], as people do not cook them properly in microwave ovens, or let them stand the required amount of time so that the heat has a chance to kill the bacteria. The issue is that cross-contamination has probably occurred at preparation (either in their manufacturing or in your own kitchen) and the heat is not enough to kill them off. Microwaves cook in a different way to conventional heat and often leave areas of the food colder than other areas so stirring and standing time are essential.

In any case, the medical approach of eliminating one organism to cure disease is becoming an outdated notion, despite media reports to the contrary.[13] An example is the flurry of medical research activity in the early 1990s around the causation of stomach ulcers by single disease-causing organisms such as Helicobacter pylori (H. pylori). It is true that the overgrowth of this particular organism can cause an imbalance leading to ulcers but unless you look at lifestyle factors too it will invariably come back. The particular pathogen is not the problem; indeed, many people have it without symptoms.[14] It is *the imbalance in the whole* that is causing one bacterium to get out of balance. We need to harness simple technologies to improve the interior ecology rather than the military type of approach of "blasting the bug" such as with broad-spectrum antibiotics. Indeed, the antibiotic approach is failing as we are seeing bacterial resistance growing. Every time you take antibiotics it gets harder and harder to reestablish your

---

12. Wood contains natural antibacterial products that protect the tree in its growth. These remain in the wood.

13. This idea has recently been translated into gene terminology; instead of the fight against an infectious organism we are now persuaded that medical advance is about eliminating the "bad gene."

14. H. pylori lives in around half the world's population and, in parts of the developing world, as many as 90% of the population carry the bug, but only a fraction of these people ever get sick. [Hamilton01].

native bacteria. There are some bacterial strains that are completely extinct in modern humans compared to native humans [Domínguez-Bello08]. This is shocking but true. Not only are we causing extinctions in our outside world, but in our inside one too. There are many factors that have caused this change, including the reduction in using fire for cooking, farming methods, and so on, but antibiotics have been the incendiary change, the results of which are permanent and deleterious in the extreme [Gillings15].

We need to *inoculate* ourselves with more healthy bacteria, not kill them. We need to get outside more, get dirty, and grow our own food (even a limited amount in pots is better than nothing). Using naturally fermented vegetables as most human cultures have done for centuries is one way of inoculating ourselves from the inside. This will be covered in more detail later. In addition, the use of natural cleaning materials rather than chemical antibacterial products is a healthier option. There are even a number of cleaning products you can use that allow you to spray with a protective biofilm[15] that naturally excludes the bad bacteria by "eating them" (in fact rebalancing the population). Moreover, the mother bacterial culture is grown continually by the manufacturer, so it is "recyclable," it never runs out and is a natural product.

Soil bacteria are important, but we have so degraded the soil, it's hard to inoculate yourself now unless you happen to live near an organic farm. However, organic gardening will help as you can create your own mini ecosystem not subject to the same destruction as our conventionally farmed soils. We are killing our environment both outside and inside. We need to begin to experience our relationship with our world much more fully in our daily lives, that is, not just through theory but by experience. We have lost this vital connection through modern city living, convenience foods, a loss of cooking skills, and myriad other ways [Pollan13]. The fact is that bacteria are all over us inside and out; we have to move away from seeing them as an enemy to be eradicated and instead to create a good balance to help us. The role of gut bacteria is vast: from controlling parasite infestation, appetite, and therefore obesity and, most controversially, neurological health and social behavior. This will be discussed this in more detail in later chapters.

Microbiome Diversity

Just like in human communities, it is the diversity of the microbiome that promotes resilience and flexibility toward stress. We need a wide balance

---

[15.] e.g., The Japanese inspired "Libby Chan" brand available in the UK and elsewhere.

to ensure our survival and health. It's important to understand we have not just bacteria but fungi, worms, protozoa, viruses, and so on in our system, all working in synchrony. But since the invention of antibiotics which have wiped out the beneficial bacteria that control the others, this balance has been significantly altered. All these other species can overgrow or establish themselves in places they are not designed to be. This balance can recover, but it takes a long time—studies have shown that it can take up to a year after a single course of antibiotics![16] Once they have overgrown it is more difficult for beneficial bacteria to manage the restoration of balance. According to GP-nutritionist Natasha Campbell-McBride, after each subsequent course of antibiotics the more difficult it gets and thus you become sick.

Another method by which our microbes remain in balance is via the production of lactic acid produced by certain "good" bacteria which is toxic to pathogenic bacteria.[17] Although many genera of bacteria produce lactic acid as a primary or secondary end-product of fermentation, the term Lactic Acid Bacteria is conventionally reserved for a genus of milk-loving bacteria of which Lactobacillus is the most commonly known. They are part of the normal flora of humans in the oral cavity, the intestinal tract, and the vagina, where they play a beneficial role.

Retroviruses

There is another type of organism – the retrovirus – that has an impact on our lives that you tend not to hear about much unless it is because of a *pathogenic* retrovirus like coronavirus. So, much of our genome is retroviral in origin; estimates range from 5–8% of our DNA was inserted by viruses. Despite the discourse and fear around COVID-19, like bacteria, viruses are neither "good" nor "bad." They can be imbalanced and in the wrong place, but they have been *essential to our evolution.*

A virus is not strictly a living organism, being just a piece of genetic information, which is able to be inserted into the host genome. So, despite the fear around them as exemplified by coronavirus pandemic, it could be said to be just a "piece of genetic information in search of a chromosome." Indeed, we could be said to be a hierarchy of DNA—the language of life— in search of greater information/ecology. Our microbes are only "bad" when the information is damaged. We should aim for greater organization/ healing ability not the demonization of particular viruses.

[16.] http://www.drdeborhmd.com/after-antibiotics
[17.] Summer Bock in a presentation for the Microbiome Summit, July 2016.

Bacterial Symbiosis and Our Evolutionary Origins

Mitochondria are the primary energy producing factories of our cells that belie our evolutionary origins. During the writing of my first book, *The Scar that Won't Heal*, I outlined, the fascinating historical basis of their bacterial symbiotic association with us– a fact that was only discovered in the 1960s by researcher Lynn Margulis, who was initially ridiculed for such a notion. About two billion years ago a unicellular organism (called Archaea) engulfed a bacterium to create the first ever multicellular (eukaryotic) organism. This is the ancestor of our cells today, with mitochondria as the remnants of that bacterium. Now this was an extraordinary revelation that doesn't seem to have filtered down to the teaching curriculum, despite having been elucidated over fifty years ago in the 1960s by cell biologist Dr. Lynn Margulis and with her husband, the more famous writer/ researcher Carl Sagan [Sagan67].

Interestingly, this event, i.e., the evolutionary acquisition of the mitochondrion, appears to have occurred once and thus spurred the radiation diversification of all of the Eucarya (organisms with a nucleus) [Lane14]. The presence of mitochondria augmented the efficiency of cellular bioenergetics by increasing the membrane surface area for the production of ATP (the energy molecule) through cellular respiration. In addition, over evolutionary history, much of the mitochondrial genome has been transferred to the nucleus of the cell, that the mitochondria energetically support the much larger nuclear genome giving it much more adaptability to its environment. So, it can be said that that the mitochondria energetically support the much larger nuclear genome.

In other words, this association between a bacterium and an animal cell allowed humans, as well as all the other Eucaryotes (multicellular animals), to evolve by giving them enhanced energy production capability and environmental flexibility. The complexity of species that we see now was only possible with the acquisition of this bacterium, as it enabled complex control and diversity of function, as well satisfying the increased energy requirements of the human brain, which, arguably, allowed us to become the dominant species. Quite something for a chance bacterial association.[18] Mitochondria is discussed in more detail in Chapter 2 as they have enormous implications for our discussion.

[18.] Or was it? Nonmaterialists may argue that this was organized by nature in some way yet unknown.

Digestion and Absorption

In order to maintain proper digestion and absorption we need the constant involvement of bacteria in our gut. Sometimes these come directly in the food we eat; in Japanese diets where they eat nori or sushi, the bacteria in these foods help break down seaweed that contains difficult-to-digest carbohydrates. These bacteria possess 60,000 carbohydrate-digesting enzymes (compared to the twenty or so that our genome has). If we eat raw foods, the plant enzymes they contain have already broken down many of the proteins, carbohydrates, and so on before we consume them, thus sparing our own enzymes from having to work so hard. So, they help in very direct ways so that we can assimilate the organic compounds our body needs.

However, the most important contribution is from our own gut flora. When the balance is wrong the gut wall becomes leaky, proteins and minerals can't be absorbed, and large protein fragments, which should be broken down, are absorbed partially digested and go into the bloodstream. This can cause immediate symptoms of headaches, rashes, gas, indigestion, and so on. But sometimes these proteins get made into complexes and it takes up to two months to produce a reaction, so you may get delayed food allergies or intolerance causing a range of symptoms like cystitis, migraines, thrush, and even cognitive problems.[19] This raises an important point as it's not always possible to easily identify the offending allergen as it can be to any food and at various timescales. So, we need to concentrate on healing the gut first rather than just eliminating foods. This is contrary to what a lot of nutritionists practice, but it is likely to change as more becomes known about keeping your gut lining intact.

Certainly, as a short-term measure in chronic conditions, it is always recommended eat a gluten- and dairy-free diet to reduce the allergenic burden on the body. However, it is not a long-term solution if your gut is still leaky, as you are not addressing the true cause. Indeed, this raises the interesting question of the issue of gluten-free living as a treatment for celiac disease. Certainly, it works but is it really a "cure"? Ideally, you should be able to heal the gut lining of most people, even celiacs possibly, but this has not been rigorously tested even though there are a few alternative doctors and practitioners out there saying it is theoretically possible. An example is

---

[19] Don't forget your brain is also part of your body, so it will affect both. Poor digestion can contribute also to depression, anxiety, and neurodegenerative disorders including brain fog, confusion, and difficulty in concentrating or staying awake.

Dr. John Douillard, a holistic Ayurvedic practitioner in the United States, who believes we can gradually put wheat and dairy back into our diet once we fix the real cause which is "poor digestion" [Douillard17]. What he means is that we don't have sufficient digestive capability (hydrochloric acid and enzymes) to break the food down properly, so the undigested protein fragments that traverse the leaky gut then trigger our immune systems to react.

This idea that it is not wheat but our guts as the problem may have some justification; after all we have been eating wheat for thousands of years—but perhaps not as currently grown. With all the hybridization and spraying, it has become a poisonous cocktail. If you make your own bread or you have a good local baker who uses organic flour, there may be an argument here. It is recommended that in any case that you *heal your gut* as a number one priority using Douillard's or many of the other protocols out there.[20]

Nutrient Synthesis

One very important fact that not many people realize is that the microbes actually make a significant amount of the vitamins and metabolites that your body needs to function. For example, an almost unheard-of vitamin, Vitamin K2 (menadione), produced by the gut flora is vitally important for blood clotting and therefore heart and cardiovascular function. If you can't handle minerals properly, especially calcium, it can form in soft tissues rather than bones and teeth—you get heart disease. Only by returning Vitamin K2 to body can we overturn osteoporosis (so no, it isn't all about drinking more milk[21]). As well as its importance in cardiovascular health, it is vital in maintaining the efficiency of your liver detoxification system via the *Cytochrome P450 enzyme* (CYP) system. In fact, recent studies have shown that administration of Vitamin K2 can help reverse poisoning by organic pesticides [Jan16] so keeping your natural levels high is important.[22] Vitamin D needs vitamin K2 to work properly (and D is vital to your health and well-being as a powerful anticarcinogen among other things). K2 comes from fermented foods[23] or high-fat soft cheeses (French cheeses) so add these to your diet.

[20.] GAPS—Gut and Psychology syndrome. www.gapsdiet.com or bodyecology.com. See also the Body Ecology Diet and GAPS protocol later in this book.

[21.] If you study populations the higher the intake of milk, the greater the incidence of osteoporosis.

[22.] However, if you have had thrombosis (abnormal blood clots in the vessels) you will be given an anti–coagulant like warfarin which inhibits vitamin K. Not good for your long-term health although it may help you recover from an operation/illness for which blood clots are a risk.

[23.] Especially a Japanese food called natto—an acquired taste as it's quite a sticky texture.

Also, the main group of B-vitamins B1, B2, B3, B6, B12, and folate, are made predominantly in the gut. B vitamins are particularly implicated in energy metabolism and mood regulation so are vital for health. Not only does the gut flora make most of your B-vitamins, but in doing so they help to compensate for genetic defects that interfere with your ability to make your own.

We vary tremendously in our ability to make vitamins; some people possess what's called a *single nucleotide polymorphism* (SNiP). This means that one of the "letters" (base pairs) of the genetic alphabet in your DNA has been swapped for another and this makes a difference to the proteins that it codes for. If you happen to have the "bad" SNiP for a particular protein enzyme called methyl-tetrahydrofolate-reductase (MTHFR—involved in B vitamin folate production), you are a termed a "poor converter."[24] This enzyme pathway is involved in both DNA production and detoxification of toxins by methylation. So, if you are slow in this area, you are likely to suffer many health problems, as detoxification and the regulation of DNA are involved in so many processes.

This SNiP also inhibits the making of a very important antioxidant called glutathione. According to Dr. Jill Carnahan, "People with MTHFR anomalies usually have low glutathione, which makes them more susceptible to stress and less tolerant to toxic exposures. Accumulation of toxins in the body and increased oxidative stress also leads to premature aging." [Carnahan13] The list of diseases that are associated with this is immense. So, it is very important to get tested if you are concerned about this as your requirements for B vitamins are likely to be higher. If you don't want to pay for a test, then try supplementing your B vitamin intake; as they are water soluble, excess is not a problem and will merely be excreted by your kidneys into your urine. There are some good formulations specifically to tackle methylation problems; they are usually much higher doses than standard formulations.

But bacteria can also come to your aid; certain bacteria make methyl folate. They can therefore prevent you developing these diseases by "mopping up" the problem. However, if your gut flora is poorly balanced then there is no mediation effect, and you are more likely to develop problems with energy and detoxification. There is even a school of thought

---

[24.] This turns out to be my issue—as I suspected. I have now been genome tested. Once I started supplementing glutathione and/or supporting my liver with milk thistle, my digestive issues have reduced considerably. It's important to work with a practitioner on this as you can overdose on glutathione. I got nosebleeds when I couldn't deal with the excess sulphur.

that says that this ability of microbes to supply missing metabolites has helped accelerate evolution by supplementing our genetic possibilities. [Stilling14]. Food for thought. Literally...

Without a good microbiome, therefore, you are likely to be deficient in B vitamins. The common symptoms are lowered immunity with constant low-grade infections: thrush, mouth ulcers, and so on. Being vegetarian for years, which was considered a healthy diet option, turned out always the case, particularly in the Western mode. As Terry Wahls, a nutritionist who healed herself from MS says, "some vegetarians do not eat meat and also do not eat vegetables, fruit, few nuts and very little protein. Instead, they are still eating sugar, white flour and little else" [Wahls16]. This was me. Eating vegetables, of course, was not enough; The reliance was on processed substitutes like textured vegetable protein (TVP which wasn't tolerated well), fungal protein (Quorn™—ditto), cheese, and pastry-based foods.[25] A very inflammatory diet in all, but, unfortunately, one that was regarded as "healthy." Regardless of whether you are vegetarian or not, if your gut flora is compromised by years of poor diet and/or antibiotics it's likely you will suffer disease at some point, usually by middle age if not sooner.

Genetic Regulation

When genetic engineering technology first became possible in the late 1980s, it was believed that humans must have upwards of 140,000 genes to account for the complexity of human biological systems. The *Human Genome Project,* launched in 1990, was conceived as an international collaboration that would begin mapping and sequencing the human genome (the sum total of the genetic material) in order to be able to find the genes for all human disease and eradicate them. At least that's what was trumpeted in the media. We all believed then that we were close to a major breakthrough in medical science. But the science was incomplete and therefore based on a false premise; that genes alone controlled the types of proteins (the building blocks of life) that were produced and thus made you who you are. However, scientists were in for a shock; when completed in 2003 it was found that the human genome contained only 23,000 genes—roughly the same as an earthworm! This has been called the *genome complexity conundrum.* The small size of the human gene pool—compared with the genomes of much simpler organisms' (e.g., rice) that has

---

[25.] All of which triggered the candida overgrowth. Candida is a fungus (yeast) and therefore is stimulated by other fungi – alcohol, vinegar, and yeast producing foods.

more than 40,000, was a very surprising finding that baffled scientists. They could not understand how you could control the many complex processes in the human body with so little of the genetic "blueprint."

What has now become apparent is "that scientists were ignoring the contribution of the bacterial microbes that make up our gut and manufacture many of the necessary proteins for us" [Galland14a]. We now know there are 2–3 million bacterial genes in your gut—so at least 100 times more than our own genes. We now know that these genes interact with ours (with some gene transfer from bacteria [McFallNgai13]), as do the protein products of the genes, so there is complex interaction between our microcosm and macrocosm. The belief is gaining ground now, that the crossover between bacterial and our genes may have been instrumental in the success of homo sapiens as a social animal [Montiel-Castro13].

In summary, we are the sum total of our microbes; if they are healthy and happy, we are. Moreover, just to blow your mind totally, there are millions of viruses that infect the bacteria called *macrophages*—which haven't even been characterized yet—and they outnumber bacteria 10 to 1. So, it has finally reached the point where even reductionist science has to concede that we are part of a larger whole. A helpful book on the microbiome, titled *I Contain Multitudes* [Yong16], was released during the writing of this current volume. The author, Ed Yong, a UK-based scientist and writer, has concluded, as we have in this book, that we are, in fact, a collection of bacteria. It's an astounding thought. Yong's title, by the way, comes from a poem by Walt Whitman, perhaps explaining its poetic touch.

It is not an overstatement that this is one of the greatest medical turnarounds in one hundred years; bacteria are our good friends, not enemies. When connected together in a harmonious ecology they have a profoundly positive effect on our health—we couldn't survive without them. They take care of us when they are in good health and provide a myriad of functions for us. They keep the gut whole and improve the function of the blood/brain barrier and thus have large beneficial effects throughout the body. They are in fact, according to Dr. Raphael Kellman, "the greatest conductor that ever existed" [Kellman16]. They can turn genes on and off which affects every system of the body simultaneously. They could be considered the "software" of the body (to use a familiar computer metaphor). They may indeed hold the master key to our life via their DNA. This understanding is so profound it shakes our worldview. The body is not a machine after all, as has been understood by science for hundreds of years; an idea that was based on the study of cadavers or isolated tissue samples. It is a living world of meaning and love, not randomness and survival.

Even "bad" bacteria like Staphylococcus aureus (much talked about as a pathogen under the moniker "staph infection") are needed for the health of the lung epithelium. It is fundamental to understand that the balance between species "plays a significant role in keeping us healthy" and bacteria are not just "bad guys" or "good guys." This is a *functional medicine* or *systems-based* model that says there is no such thing as separation of parts; everything is interconnected, much as mystics have always believed. Microbes are a device—only "bad" when the information is damaged. A greater diversity of microbial information encourages health thus is what we need to encourage, not the demonization of particular microbes in a never-ending "war" against disease.

If we follow this more environmentally/spiritually conscious line of thinking, we are led, inevitably, toward the creation of a new vision of humanity. We are certainly developing a new understanding of health and healing that acknowledges the importance of *harmony* within our bodies, and between our environment and our spirit. Seeing the soul as an important component of being human but being united with all other living things— those that live within us and those that we live among; "a vision of living beyond our small selves" means life becomes more like "painting in broad strokes."[Kellman16]. We need to abandon reductionism as a delusion of limited consciousness. We can even see disease itself is an arbitrary definition designed by humans. With a microbiome-organized view, all disease has its origin in the gut and so is the same disease manifested in different parts by the particular mind/body it inhabits. Our science needs to change to accommodate this new understanding so that selflessness can be developed as a global awareness—it seems that this may be happening finally, in microbiology at least.

Crosstalk

Of course, all these microbial guests that live symbiotically with us have to know what's going on in their immediate environment and we have to know what they're up to. Hence, there exists a bidirectional communication between gut host cells and the bacteria that live there—sometimes called *horizontal gene transfer* (HGT) —which is the swapping of microbial DNA constantly interchanging each other's genes (and now proven with our own genes too as in antibiotic resistance [Huddleston14]). This compares with the regular *vertical* gene transfer which has been understood for 300 years or so (i.e., from an organism to its offspring).

We need to have communication from our gut microbiome as to what we are taking into our bodies. In so doing, we find out what we are tolerant and allergic to. The gut mucosal epithelium feeds back information about what is present in the gut, but the bacteria are also feeding in information back to the gut. Dysregulation here sets up disease in later life—hyper-reactive people have a disrupted communication and therefore don't always know what is good for them. As an example, if they have high levels of the yeast Candida, it releases a chemical into the gut which makes people crave sugars, which of course feeds the yeast. The overgrowth of this particular species thus ensures its own survival. Marvelous for them, terrible for us.

The Gut Immune System

You may not have learnt this in school biology, but the microbiome is the organ of innate immunity. We may not have learnt that the immune system has two types: the innate and the acquired systems. When we think of the immune system, we tend to think of the *acquired* type with its exciting array of specialized cells which come to destroy pathogens (including antibodies—the only part that most people think of).

At the heart of this is the identification of foreign proteins—called "antigens"—and the production of host proteins to destroy them— "antibodies." But there is a more basic "first line" defense system (hence "innate"") in which the microbial community of the gut is highly involved. And most of us are deficient in types of bacteria that we need for this to be efficient.[26] Figure 1.2 shows how the gut is organized as an immune tissue.

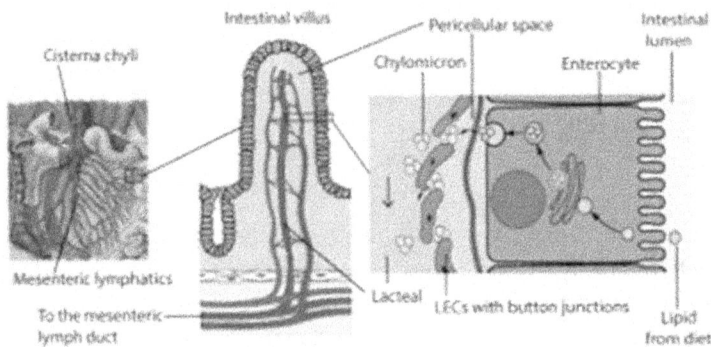

Figure 1.2. GALT (lymph) tissue.

[26.] No wonder COVID ripped through western countries while less industrialized countries, who eat food grown locally with traces of soil on, were less decimated. India and parts of Africa are cases in point.

There is a layer below the gut lining called the *gut associated lymph tissue* (GALT) that is part of innate immunity. If this is intact and working well, it can destroy any toxic bacteria that enter. But you need integrity of the mucosa (gut lining cells) to start with to make sure the barrier is working well—otherwise foreign proteins enter the bloodstream. So gut integrity is key, and the type of bacteria present has a huge influence in that—for instance Bifidobacterium species help promote integrity. When the barrier isn't working you get uncontrolled entry of foodborne bacteria and therefore proinflammatory molecules in blood and brain (the *blood-brain-barrier* [BBB] has many similarities to the gut and if one is leaky then likely so is the other).

As Dr. John Douillard, expert in the gut and immunity says, "If the intestinal skin becomes irritated by undigested food, toxins, and stress, lymph around the gut, called gut-associated lymphatic tissue (GALT), can become congested [Douillard20]. This is where 80% of the body's immunity lies and therefore has knock-on effects on infections and gut microbe balance. According to good scientific research keeping your GALT, gut epithelial cells and microbiome healthy has major implications for longevity. You need to give it the right messages and that is primarily by the kind of food you eat and how you eat it.

Gut Permeability

The intestinal mucosal epithelium (inside skin) provides the largest interface between the external environment and us and is thus a crucial regulation site of innate and adaptive immunity. We have recently discovered the existence of a complex control system that works to increase the permeability selectively; allowing gaps to open up between the cells so that certain molecules can pass through, while keeping them closed to potentially damaging molecules like food proteins (which would trigger immunity). This is mediated via a series of proteins which control permeability.

Due to the revolutionary work of a team of researchers led by Alessio Fasano in the United States, a protein called "*zonulin* was identified as one of the few known physiological mediators of *intestinal permeability*" ("leaky gut") [Wang00], [Vanustsel13]. In other words, it allows the lining cells to open up and close again to allow the passage of certain molecules needed by the body. This permeability was found to be upregulated (increased) in various immune diseases like celiac disease (CD) and type 1 diabetes (T1D) [Watts05] and so research interest was piqued because it could be a target for intervention (= drugs in most cases). Recently, however, a further discovery has shone a light on how zonulin might do this which has implications for human variability. Human zonulin is now seen as the inactive precursor for one of the two varieties of *haptoglobin* - HP); haemoglobin-binding proteins with immune-modulating properties.

Now remember, all proteins are coded by specific segments of DNA (*gene*) that you inherit from your parents (and from bacteria as we now know). The gene for HP harbors a common polymorphism (alteration of a letter in the DNA code) with two different variants: *HP1* and *HP2*. If you are unlucky enough to inherit HP2 from both parents, you will have what's called a *genotype* HP2/2. This is not good as it has been shown to be associated with the expression of different autoimmune diseases like rheumatoid arthritis (RA), lupus (SLE), type 1 diabetes inflammatory bowel diseases like ulcerative colitis (UC) and celiac disease (CD). You are more susceptible to gluten than most people and will have an uncontrolled leaky gut for three hours after eating it. You are likely to develop an autoimmune disease if you keep eating it as you will keep triggering your immune system and it will start reacting to its own tissue.

For those people with a mixed *phenotype* HP1/2 (one parent gave you the weaker HP2 gene and the other was a stronger HP1) then you will react to gluten within a short time of ingestion (though not always obviously) and should watch your intake if you do consume it all. This may correspond to those who get the lesser disorder of non-celiac gluten sensitivity (NCGS)), and is generally correlated with a state of inflammation [Daulatzai15]. HP1/1 types will have some tolerance to gluten and may not be troubled by auto-immune diseases.

Table 1. Human Haptoglobin in Genetic Inheritance and Disease

Genotype	Phenotype	Disease Tolerance
HP2/2	Leaky	Low—likely to get auto immune conditions like CD/RA/SLE/UC etc.
HP1/2	Semileaky	Moderate—but with ingested gluten may get NCGS and other conditions
HP1/1	Not leaky	Good—unlikely to get AI conditions

Maybe the reason why some people with poor phenotypes don't get autoimmune disorders is that their gut flora are well balanced from an exceptionally clean diet which helps to mitigate this poor genetic hand or it may be that they lack the other stressors like toxicity, emotional stress, and so on, all of which combine to produce disease.[27, 28]

[27.] This turning on or off of genes is termed *epigenetics.*
[28.] The HLAD-Q gene similarly codes for celiac or gluten sensitivity.

Gut-Brain Communication

For a long time, we have ignored the role of the gut in whole-body communication—we have assumed that is solely the responsibility of the brain downward to the body. It is not commonly known that there is also a complex nerve connection from the gut to the brain that signals whether you're happy or sad, the so-called "gut-brain axis." There are actually more nerve cells in the gut than in the spinal cord, roughly equivalent to the size of a cat's brain [Levine10].

So, the gut could be considered a brain too and has even been given the name the *enteric nervous system.* Most of the innervations (nerve connections) are *from the gut to the brain* and not the other way round— hence gut instinct is really quite an accurate term. The enteric nervous system probably developed first in evolution and the development of the brain agglomeration later [Naviaux16]. According to recent research, the microbiome contribution to this dialogue is significant—in fact some researchers have updated the term to the *brain-gut-microbiota axis – BGM).* Perturbations of the BGM interfere with many body systems not just limited to the gut. They can influence mood and state of mind too.[3]

Some of this may be mediated by the *vagus* nerve (tenth cranial nerve) that runs between the brainstem and the gut, as it is probably one of the most important mediators of information throughout the body; it innervates all the major internal organs and feeds messages back to the brain about your internal state. In particular, it mediates much of the *stress response,* which is highly implicated in many diseases.

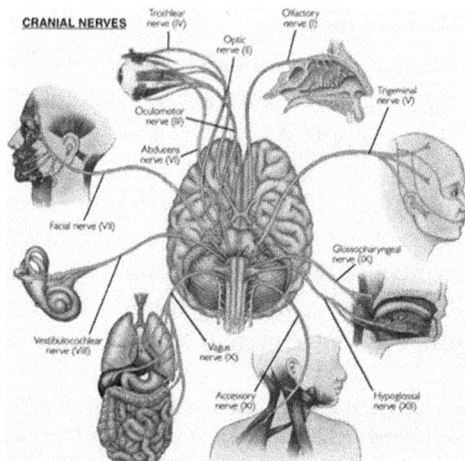

Figure 1.3. Cranial nerves.

An overactive stress response is developed from early experience (and some would say prebirth and beyond).[29] The first place you often feel this is in the gut, with poor digestion; a tendency to colic, digestive upsets and so on as a child. If this then carries on and is not addressed or remediated, then the BGM learns to be *hypersensitive to stress.* It will react even to situations that are not intrinsically threatening if the mind-body perceives the events as threat.

Don't forget, your entire life history is recorded in the unconscious parts of your brain, particularly the limbic system. Any unpleasant event where you are helpless can trigger the stress response if it has been primed by these early experiences. Many gut symptoms could be understood as a response to an unresolved trauma keeping the body "stuck" in a fight or flight state.[30] This research is described in more detail in my first book as I dealt with the issue of early life trauma and its effects on the body.

However, it is not just the vagus that dictates our complex response to the environment; the BGM axis includes the *central nervous system* (CNS), with involvement of the endocrine (hormone) and immune systems, all interacting with the huge influence of the microbiome. This two-way communication between our resident microbes is only now beginning to be identified and shows us that they are largely responsible for us having become the evolutionarily successful humans that we are. Our ability to interact socially has allowed us to learn from each other and thus innovate in ways that our less communicative cousins the apes have not quite achieved. Much of this success is due to the role of emotions mediated largely by neurotransmitters.

Neurotransmitters: Made in the Gut

It is becoming ever clearer that human behavior is affected by our resident gut bacteria. According to researcher Timothy Dinan, "the microbial ecosystem, consisting of approximately 1kg. of bacteria in the average adult (constitutes) approximately the weight of the human brain" [Dinan15]. How this control is effected is largely via the neurotransmitters that they produce. This is an exciting new area of discovery and has been described as our collective unconscious, a term coined by Jung which unites the psychological with the physical and is used in the title of Dinan's paper [Dinan15]. All you Freudian/Jungian therapists take note. In detail, he

---

[29.] This is the view of what's called "transpersonal psychology."

[30.] As Peter Levine has determined, in his fantastic book *Healing Trauma: How the Body Restores Goodness.*

believes that the importance of the microbiome comes down to three areas:

1. Gut microbes are part of the unconscious system influencing behavior.

2. Microbes majorly impact on cognitive function and fundamental behavior patterns.

3. Disorganization of the gut microbiota can negatively impact on mental health.

He posits that intervention with probiotics have a potential mental health benefit; he calls these *psychobiotics*. Most of the important neurotransmitters like serotonin and dopamine are made in the gut by our microbial helpmates; these have profound effects on our brain. According to researcher Leo Galland, "Gut bacteria directly stimulates afferent neurons of the enteric nervous system to send signals to the brain via the vagus nerve. Through these varied mechanisms, gut microbes shape the architecture of sleep and stress reactivity of the hypothalamic-pituitary-adrenal axis (HPA)" [Galland14b Abstract p. 1291].

The neurotransmitters they produce are identical to those produced by humans. We now look at each of these in turn.

Serotonin

One of the main neurotransmitters to be made in the gut or "second brain" is serotonin—the "happy" chemical. It is an astonishing fact that 95% of serotonin is made here, not in the brain, so getting your gut in balance is a key factor in alleviating depression for instance.[31] Good nutrition is something introduced to my clients very early on in our work together—but it is seldom mentioned in your average GP's surgery consultation. In point of fact, many neurotransmitters are made here in the gut. Low serotonin causes anxiety and poor digestion including leaky gut, poor motility, constipation, and so on, so the gut and brain work synergistically.

Low serotonin symptoms: rumination (over-thinking), OCD (obsessive compulsive disorder), sleep problems, and afternoon/evening cravings. Also so-called "imposter syndrome" whereby you feel a fraud in your own life and are constantly feeling you will be uncovered may be due to a lack of serotonin.

[31.] Despite what you may have been told, depression is not simply the absence of serotonin i.e. it is not a serotonin deficiency disease. Other factors like gut health, unresolved emotional experiences and genetics also play a part.

If you can recognize any of these symptoms, you are likely to be low. The standard treatment, conventionally, is pharmacological: a class of antidepressants called selective serotonin reuptake inhibitors (SSRI's). Administering SSRIs does not increase your overall level, it merely prevents the little you have being reabsorbed. It is not a long-term solution as it does nothing to address the underlying cause.

A much better solution to pharmaceuticals is foods that contain the amino-acid precursor tryptophan: eggs, turkey, walnuts, salmon, and tofu. Another solution is to use the supplement tryptophan without food; taken lingually (on the tongue) it can be absorbed immediately. The starting dose is 500mg two times a day, and then individualizing according to need. You may need to have even less if you are very sensitive to small amounts. If you can't find the powder version, then use tablets; open the pill and just sprinkle some on the tongue. However, you do need to be careful if you are using SSRIs as the two can interact. Make sure you get your GP's approval; take at least six hours on either side of your medications (if taken in the morning); so, take them in the afternoon/evening for best results.

Serotonin converts to melatonin (the sleep hormone) in the body and this can affect sleep substantially. So, if you find, that sleep is affected by your depression, then add melatonin sublingually to go to sleep and perhaps take a slow-release formulation to stay asleep. It helps gut motility too, and that relieves constipation should it be an issue for you. Apart from its ability to help sleep problems, melatonin is also a great antioxidant; and helps with cancer prevention. It is a highly underrated nutrient that is beneficial for many diseases.

Dopamine

Dopamine is our reward neurotransmitter which drives us to seek and find pleasure. It also helps regulate movement and emotional responses toward that goal and can drive addiction if it is not balanced. Deficiency is found in people who have Parkinson's Disease (PD), an illness that can cause movements to be uncoordinated and spasmodic. Since 50% of dopamine is made in the gut by our microbes, it is really important to have a good balance.

Low dopamine symptoms: ADHD,[32] depression, restless legs syndrome, inability to handle stress, fatigue, mood swings, inability to concentrate,

---

[32] Currently ADHD in children is treated with Ritalin—a powerful Dopamine-boosting drug that has serious side effects including growth stunting and appetite suppression. How much better it would be to get their gut flora right.

forgetfulness, and a failure to finish tasks. However, in adults there is a lot of crossover with low thyroid and adrenal function so, if suspected, a simple blood or urine test is recommended. The precursor of dopamine is tyrosine. Foods that contain high tyrosine include eggs, bananas, dark chocolate, avocados, chicken, and beef. Where dopamine is low, serotonin imbalance is often found, so the foods that enhance serotonin (i.e., contain tryptophan) are also important.

GABA

Gamma-aminobutyric acid (GABA) is another very important neurotransmitter; the major inhibitory (calming) neurotransmitter of the central nervous system. There are extensive GABA receptors in the muscles, so it helps them maintain relaxation and supports the endocrine (hormonal) system.

Low GABA symptoms: Physical anxiety, loss of voice caused by spasms in the throat, tightness in the shoulder muscles and gut, panic attacks, cravings, and stress eating, for example, that feeling that you must have a glass of wine/or cravings for sweets. Therefore, if your anxiety symptoms feel more physical or you suffer from the food cravings rather than low mood and sleep issues, this could be the more problematic neurotransmitter for you. It therefore forms an obvious target for intervention.

It can be reversed by administering the amino acid GABA on the tongue. GABA is a really key amino acid for relaxation and so you would be wise to treat your gut with care and love as it helps to make you feel good. Luckily there is a bacterium that makes this; lactobacillus rhamnosus. When increased by inoculation (supplementation usually), it changes GABA expression in the brain and reduces anxiety but interestingly, when the vagus nerve is severed this reducing effect is not seen. This shows that most of what is going on in the gut is directed to the brain via this nerve. GABA is supposed to not cross the *blood brain barrier* (BBB) but there is a theory that the vagus nerve and gut/enteric brain may circumvent the BBB [Boonstra15]. Some molecules may even be able to travel along the nerve itself. Remember if you have a leaky gut then you are likely to have a leaky BBB too as they are similar tissues so it may be this is the issue for you. So, try it and if you feel relief then you clearly need more of it.

But what of the influence of the gut bacteria? Lacticaseibacillus rhamnosus has been associated with GABA and recently new research, identifying the 50%–80% of the microbiome that remains to be identified, has highlighted a GABA-eating bacteria: KLE1738 [Coghlan16]. Clearly if this bacterium is overabundant, it is problematic. But it is not yet clear whether this is cause or effect. Trudy Scott has written extensively on this particular neurotransmitter, for more information refer to her website.[33]

Glutamine

Glutamine is also worth considering as a supplemental amino acid taken orally to help stabilize blood sugar, heal the gut, and calm anxiety. Some glutamine converts to GABA—the main calming neurotransmitter in the body. Again, there is the problem that it cannot traverse the blood brain barrier (BBB), so oral suspensions are argued not to work. However, in practice, it seems they do as it is absorbed by the enteric nerves and traversed up to the brain that way. This would explain the gap between theory and practice. Also, it is recycled in the glial (helper) cells of the brain, thus supplying the neurons there too. According to studies, as reported by stress consultant Duane Law, "As we get older it gets harder and harder for the brain to form the enzyme necessary to make GABA from glutamic acid; manganese can correct this" [NSC00].

Low glutamine symptoms: constant infections, fatigue, bowel changes, loss in muscle mass, and kidney problems. With clients, it is suggested that they start with this one as it is so fail-safe. As an added bonus, it seems to stop cravings in their tracks too. Some doctors are of the opinion it is not wise to take it if you have had cancer, but the research does not support this as it is highly cancer-preventing.

Histamine

Histamine is a biochemical "alarm" agent released by specific white blood cells called mast cells in the body, which attract other parts of the immune system to come and attack a foreign body such as a toxin or bacterium entering through a cut or injury. "Mast cells interact directly with bacteria and appear to play a vital role in host defense against pathogens" [Amin12, Abstract p. 9]. Some people have unstable mast cells (due to genetic predisposition), which can be triggered by sunlight, chemical agents, vibration, bacteria, parasites, metals, medications, and anything that's not supposed to be in the body. This causes an overproduction of histamine which, if there is not enough enzymatic degradation via the breakdown enzyme diamine oxidase (DAO), results in a high level in the blood.

---

[33] http://www.everywomanover29.com/blog/anxiety-summit-microbes-gut-psychobiotics-potential-treatment-anxiety-depression/

A good clue is if you flush easily with alcohol or are an "angry drunk" then you are hypersensitive to histamine release in your brain. Alcohol blocks DAO and thus allows more histamine to hang around. This affects other neurotransmitters like dopamine too. It is a double whammy of neurotransmitter imbalance in the brain, and therefore it is best to avoid excess alcohol in this situation and get your gut balanced. Some people are more sensitive to histamine levels; they exhibit flushing, shaking, and even anorexia; they find they can't gain weight. Multiple chemical sensitivity (MCS) can be the end result of such histamine instability. It is a combination of internal and external triggers, including trauma. "Mast cells are ubiquitous in the body especially in tissues exposed to the outside, such as the skin, intestine, and lungs, including the brain, the richest of which is the hypothalamus [Theoharides13]. You will remember that the hypothalamus is the part of the brain that responds to stress (the "H" of the hypothalamus-pituitary -adrenal HPA axis). This may be highly significant.

Histamine release prevents recycling of cell organelles (the tiny components of cells including mitochondria) that reduces your energy levels. This is discussed later on in the book in terms of mitochondrial function. Mast cells are involved in cancer growth too [Maciel15], so getting them under control is important for long-term health. For instance, triple negative breast cancer (the one that's so difficult to control) can be treated with antihistamines so you can see the link.

Most people are aware of histamine as they may take "antihistamines" to combat allergies. But you need to be careful with antihistamines that can unbalance the body; there are four different receptors, and you need a good balance of these four neurotransmitters that is thrown out by these powerful drugs. Better in the longer term, is to avoid high-histamine foods; alcohol, balsamic vinegar, Chinese take-out (with MSG), mushrooms, chocolate, and tomatoes. I, for one, always know if there is MSG in food that I've eaten; first I get the postprandial "high" and the next day the most awful headache. Migraines are linked to histamine release too, so it is not surprising that headaches are a response and that migraine sufferers are likely to be imbalanced. Gluten and milk casein also cause non-allergy mediated histamine release, and are problematic for those who are sensitive to histamine. There is some indication that people with pale skin, red hair and/or freckles appear to be more likely to be affected by gluten and casein. This is because the genes that cause red hair and freckles affect the endorphin production/receptor system so individuals with redheads tend to require larger quantities of anaesthesia (around 20% more) but experience increased analgesia to morphine which may affect tolerance levels to the opioid-like peptides in gluten and casein.

There are other less obvious ways to lower histamine: meditation and guided visualizations, for instance. Researcher Yasmina Yekelenstam believes it is via this effect that histamine reduces the stress reaction. Histamine is a necessary part of your inflammation reaction; however, it's the *overproduction* that is the problem. Stress is the biggest trigger, so if you get stressed about how you are feeling or have chronic stress (which can become a conditioned response if you have suffered childhood trauma of any kind), then you need to get some psychological/ emotional release therapy.

Cortisol

Cortisol is a stress hormone produced by the adrenal glands (the outer part or cortex, hence the name). It is produced as a short-term antidote to insulin and adrenaline to conserve body resources. However, if you are continually stressed, as many people are, cortisol levels remain high. In some people this becomes adrenal fatigue (the adrenals stop being able to produce it) and cortisol levels actually drop. In all cases, cortisol levels are altered from their normal diurnal variation; high in the morning (to get you up) and low in the evening (so you can sleep). Some people have day and night reversal, therefore, sleeping in the day and wide awake at night. Melatonin follows it inversely.

High cortisol can *cause* anxiety as well, so GABA (the calming neurotransmitter) won't work if you have this type of anxiety. Low cortisol means you simply can't get going—particularly in the morning. If your energy levels are low at this time, then low cortisol might be the problem. This can be checked very simply with a saliva test available online or through the mail.

You must look at stress first, as this is the root cause. However, the definition of stress here is very broad; it could be life stress, gluten sensitivity, parasites, etc. You need to identify your particular combination and take steps to reduce it. One method that has been largely ignored is through the medium of probiotics and altering the gut flora. *Prebiotic* intake, such as galacto-oligosaccharides (a specific carbohydrate that gut bacteria favor), has been shown to lower high cortisol too—this reduces anxiety and stress by getting the gut in balance. For example, taking phosophotidyl-serine (Serophos) two hours before your peak can be helpful.

There are other products from plants like magnolia and philodendron that are said to have restorative effects by binding to the cortisol receptors and, unlike their pharmaceutical counterparts, do not cause drowsiness. There is some evidence of the health benefits of raw milk (unpasteurized) that may be due to its content of probiotic beneficial bacteria such as Lactobacillus species, plus the digestive enzymes and immunoglobulins (antibodies) it contains. These are all destroyed by pasteurization (heating to high temperatures).[34] So, in summary, all of these methods can work in synergy, so it is something best approached with the help of a qualified nutritional therapist to work out an individual plan.

Neuropeptides

Besides specific neurotransmitters, there is also a complex dance of small molecules called *neuropeptides* that communicate emotions between the gut and the rest of the body—two of the most well-known are vasopressin (secreted by the pituitary to control salt concentration in the blood—and therefore regulate blood pressure) and oxytocin (the "love hormone" as it is produced around childbirth to promote bonding). Without going into full detail here, you are referred to the book *Molecules of Emotion* by Candace Pert' [Pert99]. These molecules constitute a general communication system that gives a finer detail of emotional response. It is highly likely that gut microbes respond to these signals as well. Think of the connection as being a finely tuned web rather than a simplistic linear cause and effect relationship.

Stress and the Limbic System

Stress causes direct effects on the body via a cascade of hormones that occurs when the stress cycle is induced. Physiologically, stress is any event that takes the body out of balance or *homeostasis.* The parts of the brain that we particularly focus on in the study of the effect of trauma (highly significant emotional stress) are contained within the limbic system or emotional brain. These consist of the hippocampus (memory center), insula (emotional center), and parts of the prefrontal cortex (PFC—your master controller) all of which shrink; and the amygdala and hypothalamus that

---

[34.] I have recently discovered a local supplier of raw milk (it is banned in the United States completely!) and, after years of considering myself lactose intolerant, I can report I get none of these issues with raw milk. In fact, it feels really good on my gut. It is not available generally though, so you need to carefully check out your local area.

grow after trauma.[35] Most of these parts of the brain are unconscious, hence the stress itself is largely so. You will not necessarily know you have been triggered by events in your life, but your mind and body respond anyway with this preprogrammed stress response. Hence it is highly important to reverse the unresolved emotions, memories, and belief systems around these events, so that we are able to reverse these negative adaptations of the brain. In so doing we encourage specific neuron growth in areas such as the hippocampus that allow improvement in memory and so on. *Reversing the message of threat* of unhealed emotions is an important precursor to healing that every person would be well advised to address.

The Limbic System

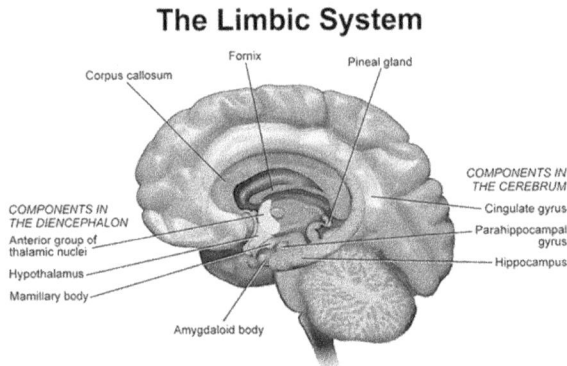

Figure 1.4. The limbic system.

The final mechanism by which stress affects the gut is via changes in permeability of the gut lining. This is also a side effect of continual high cortisol levels from chronic stress as mentioned, but there may also be a direct effect. What is clear is that stress has major implications for the production of inflammatory chemicals in the brain and body, leading to a *feed forward cycle* of heightened stress, poor resilience and mental health. Getting the stress response down to manageable levels is a major goal of most therapeutic treatment. If you ignore this, it is unlikely to be successful in the long term.

Disruptors: Glyphosate

There are many ways in which our gut flora communicates with us as we've now seen—and many ways in which this complex web can be disrupted. One of the most insidious and increasingly common is the weed killer

---

[35.] In fact, depending on which part of the autonomic nervous system is being triggered, sympathetic fight and flight versus parasympathetic rest and digest, these areas can change in size dynamically over a few days.

(herbicide) glyphosate. As a professional gardener with a long history of chemical gardening, I can remember blithely telling my clients that glyphosate was perfectly safe as it "denatured" (was rendered safe) in the ground and did not persist in the environment. How wrong I was! Or how misled I should say, as it was the propaganda of the chemical companies at the time that we all assumed was correct. In fact, only 15%–20% leaves via the roots and the rest is stored in the plant (i.e., the leaves and stalks—the food component that we eat).

Glyphosate (marketed as Roundup) usage has grown exponentially since the time I first became aware of it—and shockingly we now have *no choice whether or not we consume it*. If you look up the definition online, you will see it is not just an herbicide but a "crop desiccant." This is because not only is it used on genetically modified foods (GMOs) but, in most Western industrial nations, it is routinely used to dry out the crop before harvesting, as it is cheaper than drying the crop manually, that is, in barns or storage sheds. This has disastrous consequences for our health as this "process that results in contamination of non-GMO grains, (is) one of the main exposure routes in the EU (European Union) where GMO crops are not commonly grown" (and is the normal exposure route in the United States too) [SIS15].

Glyphosate is not an inert chemical that only kills selective weeds. Remember, it does that by interfering with a major metabolic pathway called the Shikimate pathway,[36] which is a pathway by which plants make proteins for growth.[37] The plant then, unable to make necessary proteins, dies. However, this pathway is also shared by bacteria, fungi, and the bodies of insect pests, so it disrupts the whole ecosystem that the plant relies on to grow. That's why the crop plant then has to have a resistant gene spliced into it to make it resistant to applied herbicides which then only affects "weed" plants around it. Now note that if it is affecting bacteria and fungi in the soil, it is also going to affect us when we eat the food grown in that soil and especially when we consume the plants themselves. It has disastrous effects on our microbiome [Samsel13]. It also increases the permeability of our gut lining (aka leaky gut) due to the production of a protein called zonulin so that we lose selective permeability and anything can come through, good or bad. Undigested food proteins are then able to permeate through and this

---

[36.] A seven-step metabolic route used by bacteria, fungi, algae, some protozoan parasites, and plants for the biosynthesis of folates and aromatic amino acids (phenylalanine, tyrosine, and tryptophan).

[37.] Also by which we make a lot of our neurotransmitters; hence glyphosate disrupts our neuroendocrine system.

causes an immune response of the body to our undigested food and sets the stage for autoimmune diseases like Hashimoto's, rheumatoid arthritis, and celiac disease which not surprisingly are rising exponentially.

It disrupts digestion directly by suppressing the digestive enzymes cholecystokinin (CCK) from the pancreas and affects the cytochrome P450 detoxification pathway in the liver so that we can't even detoxify our blood and cellular fluid properly. It even prevents production tryptophan that is used by the body to form important hormones like melatonin and serotonin as we have just seen. This can cause problems with sleep and mood regulation.

Glyphosate is also an endocrine disruptor; rats which were fed genetically modified soy developed issues with their kidneys, and eventually their ovaries and testes—it affected sex differentiation in short. Males would end up with female characteristics. By the third generation, hamsters had become sterile. You can be sure that if this stuff is happening in lab animals, which are considerably smaller than us, then undoubtedly these things are happening in our guts too. It is worth wondering incidentally if the rise in "gender dysphoria" in children is linked to these hormonal changes. We know that the male brain is formed by being bathed in testosterone in the womb [Brizendine17], and the female is the default gender due to absence of that hormone. Perhaps if there is hormone disruption of the microbiome it causes a mismatch between the sex of the brain and the body. Certainly, there are autistic symptoms that have been attributed to problems with such hormone disruption, as well as from differences in brain function generally and a lot of crossover in gender dysphoria and autism, especially in young women. For a summary, check out the work of Norman Geshwind and Simon Baron Cohen [HDC00].

Not only that, but in the United States and other countries who have bought the promise of high yielding crops via the introduction of genetically modified organisms (GMOs), there is an even more dangerous risk. Since the advent of genetic engineering, we have found ways to splice bits of foreign genes into the genes of unrelated organisms without any biological controls provided by normal mutations (some of the offspring of which die due to the mutation being unviable). In the case of genetically modified plants, parts of a bacterial gene are inserted into the plant's genes that make the plant resistant to herbicide to become so called "Roundup ready" plants like soy, corn, and the like. This means the crops can be liberally sprayed with Roundup to kill the competing "weeds" that potentially reduce the crop (but that also bring in beneficial insects and add to the organic layer of soil). If you only look at crop yields in terms of bulk produced per acre of land, then certainly they are increased by this technology. But there are major changes in the soil microbiome from the lack of natural aeration and the permanently altered genes of the decaying weeds which then cross-pollinate with natural (unmodified) weeds and plant life. There is a more insidious threat too from the bacterial genes themselves which may also be spliced into our gut flora when we consume these foods (bacteria swap genes much more readily than we do). We simply do not know yet how often this occurs. But we do know that agricultural workers who farm the land in developing countries (and therefore use less machinery and come into contact with the chemical directly) seem to have a much higher incidence of disease.

Glyphosate has, in fact, been classed as Class 2A carcinogen by the World Health Organization (WHO) recently [Guyton15]. This means it is "probably carcinogenic" to humans based on the limited evidence we have at the moment. Of course, this has been jumped on by the alternative health community and those seeking to get GMOs banned and this has then been vigorously defended by the chemical industry (Monsanto, makers of Roundup, in particular[38]). Doubt has been cast on this decision recently when a further report suggested there is no evidence of cancer causation through ingestion [Nelson16]. However, a recent review has gone so far as to be called a "statement of concern" because it says the science is now pointing very clearly to a serious problem in terms of rising incidence of exposure and inadequate testing and setting of statutory safe levels [Myers16].

Glyphosate is also a *chelator* (it binds minerals) like cobalt, magnesium, iron, zinc, and especially selenium (Se), which is an important mineral for the thyroid. Selenium enables the conversion of thyroid hormone T4 to the active hormone T3. Without it, conversion remains poor, and metabolism slows down. The thyroid is a regulator of energy and without enough T3 we can't make enough energy for many of our functions. This might explain the epidemic of thyroid hormone diseases like Hashimoto's thyroiditis (more on that later).

If that weren't enough it is also a very powerful broad-spectrum antibiotic; it kills mostly the beneficial bacteria in our gut which keep the more pathogenic species in check and it leaves untouched the pathogenic ones like E. coli, Salmonella species, and so on, to wreak havoc in our guts.

[38.] Monsanto recently changed their name to Bayer it seems to avoid past controversies.

We have already seen the results in our farm animals who are getting sicker due to mineral and vitamin deficiencies, requiring more antibiotic use to compensate. We then ingest this when we eat meat; another reason to make sure to eat organic (if we eat meat at all).

Glyphosate accumulates in the atmosphere, being found in groundwater[39] and sea water [Mercurio14], and *bioaccumulates* (gets more concentrated) in humans who drink that water. They then pass it on to their offspring in breast milk when they feed their babies [Honeycutt], [Bus15].

This is bad news as it is endemic in food production especially in the United States, where the genes have undoubtedly contaminated non-GMO crops (as there is no way to control the genetic activity of soil and gut microbes). In the United States, there is a proportionate rise in obesity and chronic disease that may result from this change in the genetic code of the gut flora. Here in the UK, we are not far behind. The evidence from use in the developing world is that it is causing all sorts of illnesses among agricultural workers who directly inhale or touch it, but this is only the "canary in a coal mine" in terms of long term effects. We won't link changes in the microbiome to such indiscriminate use until successive generations of children and adults who have been exposed then go on to have more children with altered microbiomes. But the signs are already with us.[40]

Many in the scientific press are just as bemused, and current reporting struggles to really deal with the issue as it is so politically loaded [Cressey15]. This is due, in large part, to the lack of scientific understanding in the general public and media who simply reduce things down to a few tabloid headlines that do nothing to further informed debate. They are missing an important piece of the puzzle that the problems are not just through the *direct* effects of ingestion on our human cells (as a lot of the studies have used human cell cultures not real human subjects), but on our gut and soil bacteria—the earth's microbiome. The jury is out as to the effect on human health at the moment, but cancer takes many years to develop whereas other diseases are quicker to manifest but likely multifactorial factors contribute.

We are only at the start of understanding the long-term effects of such unmitigated meddling with systems we barely understand. So, for now, whatever you do, try to minimize your contact with glyphosate: eat organic and avoid home use of weed killers containing glyphosate.

[39.] This directly contradicts the official line that it is neutralized upon contact with soil.

[40.] An exponential rise in autism, anxiety, behavioral difficulties like ADHD, obesity, etc. is already evident. All are related to neuroendocrine disruption.

The Microbiome of the Mouth

So far, we have only looked at the microbiome of the gut—but there are many different microbiomes in the body with their own particular diversity of species. Every one microbiome is slightly different.

You will remember from the introduction that the oral cavity has a very different make up to the gut. Dentistry needs to acknowledge the importance of the microbiome within the mouth. Currently only holistic "biological" dentistry does this. The mouth is the portal to our overall health. A vast majority (80%) of people over 30 have some stage of gum disease from gingivitis to advanced periodontitis. Gum disease is extremely common and tolerated as "normal." It is not. What goes on in your mouth has a profound influence in your general health; poor oral health, particularly gum disease, increases your risk of heart attack and likely death. Astonishingly, gum disease is a higher risk factor than elevated cholesterol for heart disease although they are linked [Vedin14]. Moreover, this is true for stroke, too, but we never hear this from our dentist or other medical professionals. If you have gum disease you have ten times greater the risk of heart attack and seven times the risk of diabetes. If your gums bleed, this is not because the brush is too hard (a common explanation); it is a sign of chronic low-grade inflammation. Joint pain is also a common sign. But you may not feel a thing until damage is well underway, so it is a silent killer. One way to check whether you suffer from this is to have a blood test of CRP (C-reactive protein) which should be part of a standard blood test but sadly is not considered often until it's too late.[41]

So, we now know that the microbiome of the mouth, being so near the brain, is important. We have been long aware of bacteria in the mouth; we have deployed ever more sophisticated techniques to eradicate them (assuming them to be all "bad"). This has somehow been linked with plaque formation as if it is solely the cause of the problem. However, as ever, it is an incomplete picture. Plaque is simply an unhealthy expression of bad bacterial balance, *not* the problem itself. If we destroyed it completely, we would die, as we would be unprotected. Plaque is part of the biofilm that protects our mouth from invaders. The best way to counteract a poor balance in the mouth, next to diet, is by using prebiotic toothpaste like Revitin. This restores the microbial homeostasis (balance) and helps to "prune" the bad bacteria. Available online and in stores in Europe and the US. And in

---

[41.] But certainly, if your gums bleed regularly on brushing you need to take action—and that is not about mouth washes to "kill" the bacteria. It is about better dental hygiene and diet.

the meantime, there are always online stores. Rinsing with coconut oil in the morning is a simple way we can modulate our oral bacteria, but it is distasteful to some.[42]

Tooth Decay

Tooth decay is another sign that something has gone wrong in the balance of bacteria in the mouth, most commonly linked to poor diet, (too many carbohydrates, especially sugar) and *not* inadequate brushing as most dentists and toothpaste manufacturers would have you believe. Dr. Weston Price, a famous dentist in the 1930s, studied this link with diet around the world and found that once native people adopted a Western diet their dental health worsened considerably [Price09]. His work is carried on by the Weston Price foundation in the United States. It isn't just the health of the teeth that are affected but the jawbone too—they are both made of similar substance—bone and dentine share many similarities. People on high sugar diets often have poor dental arches and crowded teeth. My childhood was a symphony of sugar addiction and constant dental surgery as a result.

So, what is the relationship of the microbiome in the mouth to that in the gut? There are many similarities, but the main difference is that the mouth has exposure to the external environment, so it has a primary protective role. As such, it is important not to disturb that balance with either regular mouth washes (which usually contain the chemical chlorhexidine or natural antibacterial oils like tea-tree and peppermint, both of which disturb the balance—use saltwater or coconut oil instead. Surprisingly, peppermint is often used to flavor toothpastes, so, you might think that adding peppermint oil to your own home-made toothpaste would be a good idea. But it seems you have to be careful about which herbs to use—aloe vera, thyme, and fennel have all been found to be safe and effective and are better additions.

You need to boost the oral ecology with substances that promote your oral ecology; we could almost liken it to "gardening of the mouth." Remember, toothpaste was invented by soap manufacturers—if you look at most of the commercially available ones most contain an emulsifier such as sodium lauryl sulphate (SLS) which is an industrial surfactant (binding agent for oils, also used in shower and bath gels). This is not good for you or your mouth. Most also contain fluoride, which is supposed to help gum disease. However, for adults the evidence is patchy at best, and it may

---

[42.] Oil pulling, as it is called, takes out toxins from the gums and removes pathogenic bacteria.

indeed lead to overdose, given that it's also in the water system in most western nations. There are also concerns that it may weaken children's bones with higher fracture rates in US states where there is a higher level of fluoridation [Lindsay23]. Another highly contentious arena, heavily defended by vested interests!

It is recommended to discard the commercial toothpastes and buy natural SLS/fluoride-free ones or use your own mixture. For a really good clean, try oil pulling using sesame or coconut oil. You put a teaspoon of oil in the mouth first thing in the morning and swish it around the mouth for a couple of minutes before spitting out. This removes some of the build-up of overnight bacteria and prevents them from going down into your gut when you swallow. You must do this before you eat or drink anything for it to be the most effective. The oil pulls out not just the bacteria from between the teeth, but it also pulls heavy metals from the gums too. Coconut oil tastes the best, but you must be careful not to spit it down the sink as it can solidify later (it liquefies at body temperature but is solid normally).

So, what is the answer here? One simple thing; *replace detergents with nutrients*. Your saliva is your best defense against tooth decay and it should be alkaline. When you eat acidic foods (meat, eggs, bread, etc.) and not enough alkaline ones (leafy vegetables and seeds), your saliva is simply not alkaline enough to remineralize the teeth as it is designed to do. This is probably the biggest reason that tooth decay occurs. You can follow a dietary program to remineralize your teeth that includes highly nutrient dense foods (usually vegetable protein powders like hemp, spirulina, pea, etc.) added to breakfast smoothies and juices, with a 60/40 veg/meat balance for other meals of the day. Of course, if you are vegetarian, you will not eat meat, but the mistake many veggies make, is to substitute carbs including wheat, dairy, and such for the missing protein. Having switched to a semi-veggie/"pescatarian" diet (fish and occasionally some organic chicken cooked on the bone) suits me better. Note I am not saying everyone should do what I do, but for me there was the realization that I was not receiving the right balance and needed to add more protein. How you do that is a personal choice.[43]

It was many years before I learned that tooth decay itself is reversible; that's right—you can heal it. Until recently the only recourse was to go the dentist and get the tooth filled or extracted, but there is now more

---

[43.] Nutrient requirements/food tolerance may be determined by your blood type which is linked to your hereditary genetics. Get your genome tested if in doubt.

information on preventative dentistry that talks not just about preventing further cavities, but *reversing ones that you already have* (albeit not possible if you already have a filling in it). Now there are a number of good books on this subject that talk about this, so if you have never explored this idea, check them out [Halvorsen10], [Nagel10].

Nutrients for the Mouth

Mineral deficiencies, like magnesium, can weaken bones and teeth, so we need to make sure our diet contains enough and if not (or we have overt symptoms like tooth decay) we need to supplement with "food state" vitamins and minerals (that is they are as close to the real thing as possible, together with the necessary cofactors that make them more active).[44]

The most important nutrients for the mouth are

- magnesium

- vitamin C

- ubiquinone (oxidized form of CoQ10)

- MSM (Methylsulfonylmethane)

- vitamin A

Effects of Fluoride

Much has been said about fluoride, both positive and negative. Fluorine is not intrinsically a "bad" element, but it is most often introduced as sodium fluoride (NaF) which is not so good. Both teeth and bones are naturally made of hydroxyapatite, a composite mineral of calcium and phosphate with hydroxyl (OH) groups attached. When these are replaced by fluoride to form fluoro-apatite, the resulting teeth become very hard as it has no flexibility. So, not surprisingly we can get brittle bones and teeth, because it is used long-term and in an uncontrolled way. We do not know the result of dosing everyone, regardless of age, ethnicity, and gender identically via water fluoridation. We suspect that women generally have lower tolerance to fluoride[45] and that it affects dark-skinned people more than light-skinned—something that is usually hidden in the back of reports.

[44.] Food state nutrients are generally made more bioavailable than synthetic vitamins. For example, cherry extract contains a large amount of vitamin C but with bioflavonoids which makes it more bioavailable to the body.

[45.] Possibly this is because they are more prone to low thyroid to which it has been linked.

Effects of Medication

More than 600 medications promote tooth decay by inhibiting saliva including, most importantly anti-gastro-esophageal reflux disease (GERD) medications like proton pump inhibitors (PPIs) and nonsteroidal anti-inflammatory medications (NSAIDs) like ibuprofen/Advil). Avoid them if at all possible. There are other ways of reversing the conditions these medications are supposed to treat; most potently balancing the gut flora and eating nutrient dense, alkaline foods.

The oral microbiome is as unique as your thumbprint. It is hard to introduce new gut bacteria directly into the mouth, we need to look after what we have and allow the right conditions for the "good" ones to survive. This is perhaps even more important than with the gut flora. The mouth is the top of your gut, after all, so it only reflects what is going on below.

The problem is the medical model has kept dentistry and medicine very separate and dentistry is not seen as a medical specialty at all. But it should be, as the same processes are at work; chronic inflammation lies at the heart of all of these diseases and possibly more than we yet know. It is strongly suspected that most chronic disease has the same processes underlying it and the expression in one part of the body or another is due to your weaknesses/genetic/metaphysical factors.

The Microbiome of the Skin

Skin is another organ of the body; some would argue it is your largest. There are many people who think that this is an "old wives' tale." Far from it. The skin allows you to absorb things from the environment (hence nicotine patches, etc.) as well as protect your body. If your liver or colon is blocked or toxic, your skin will show rashes or other issues as your body attempts to get rid of the toxicity through the skin. There are a range of common skin problems like psoriasis and eczema that are more common signs of an imbalance. Inflammation lies at the heart of these diseases too.

Just below the surface layer of the *skin associated lymph tissue* (SALT), which is full of T-cells (part of your immune system – analogous to the GALT in the gut). There are two branches of your immune system: the one that we are born with (innate) and the adaptive (learned) immune system. T-cells are part of the adaptive part. But there exists a type of cell that helps to communicate between them called *dendritic cells*, and they "process antigen material and present it on the cell surface to the T cells. They act as messengers between the innate and the adaptive immune systems" [Wikipedia 2] and are intimately communicating with the microbiome. Intolerance or tolerance of foods begins here. If you are intolerant to a food it leads to the expression of allergy in both the gut and skin—remember the gut is just your internal skin, they form from similar tissues in embryonic development. So, your skin actually regulates development of allergy, along with contributions from your genetics, epigenetics, toxins, nutrient status, and so on.

We actually have similar bugs on our skin as in our gut (though there are significant differences). The bacterial skin and gut phyla (genetic groups) share similarities but there are also differences because skin flora is exposed to air in different amounts. Some live deep in the dermal layer not just on the surface. *Commensal bacteria*[46] (those that live with us in symbiotic relationship) go deep into the dermis helping to regulate the immune system. When there is a wound, it is the bacteria here that help recede the surface area and thus heal the wound. Interesting information when we consider we like to make the wound sterile in most cases![47]

The skin flora also help to reduce the signs of aging too, so it behooves us to keep them happy. Fatty acids are so often depleted with aeing of the skin, and in conditions such as eczema. The balance is usually too many rigid processed fats instead of the flexible "good fats" from natural cold-pressed oils, and such. A diet rich in these will do a lot to keep the skin healthy. Topical application (directly onto the skin and hair) with coconut or sesame oil[48] can make quite a difference. A suggestion is to use sesame oil when going out into the sun and coconut oil for general use after showering (although you only need a tiny amount as it can be quite greasy).

How we treat our skin affects our immune reactivity in very practical ways; allergy is primed by over-washing, over exposure to antibiotics, antibacterial hand-wash, toxic shampoos, shower gels, lotions, and incidental exposure to cleaning products. All these compounds are cumulative and damaging to the skin microbiome. If you consider the skin the largest organ in the body and the fact that up to 60% of skin care products goes into your bloodstream that way, it becomes a major source of toxicity. There are certain chemicals that are particularly dangerous; avoid anything with triclosan in it as this is a registered pesticide and in a lot of hand sanitizers and so on. Avoid coal

---

46. The word commensal is derived from the Latin "cum mensa" or eating together.

47. The lungs also are not sterile as previously believed.

48. Sesame oil actually contains as a natural UVB inhibitor to acts as a sunscreen too—without all the nasty chemicals.

tar soaps and antidandruff shampoos as they contain a lot of chemicals. The problem is a lot of the offending products can be listed as unidentifiable code names such as FD & C6 which make identification difficult. Get to know your chemicals (and don't be deceived by "natural"-sounding products that are expertly marketed to trick people into buying them – so-called "green-washing").

One of the worst offenders is DEA (diethanolamine—an industrial foaming agent used in shampoos), which has been linked to bladder and esophageal cancer and formaldehyde (in nail products) which kills bacteria, good and bad (hence its use in preserving bodies). But there are many others such as mineral oils (i.e., petroleum products which can cause birth defects—check out baby oil for instance[49]) lauryl sulphates like sodium lauryl sulphate (SLS), phthalates (associated with reproductive problems and liver cancer in animal studies), parabens, propylene glycol, and so on. Even talcum powder is not safe, having been linked to breast cancer.[50] The most dangerous above all, as far as I'm concerned though, are underarm deodorants containing aluminum (that blocks the pores stopping you from sweating). This is not only unnatural but, being applied under the arm, is also absorbed in a part of the body the closest to the breast lymph tissue. We are told to avoid aluminum in cooking pans, so how come we can safely apply it to our body? Use a rock-salt roll-on or natural deodorant instead. Check the label to make sure it is aluminum-free.

Even the fragrance and perfume counter is chock full of artificial chemical synthetic scents that cross the blood brain barrier and cause unknown problems. Possibly by now you are beginning to feel overwhelmed—it defies belief that these "beauty" products can be so damaging to us. They are particularly bad for growing children. A recent report found that young women actively using makeup and body care products daily tested positive for many of these chemicals in their urine [CSC00]. That some of these are carcinogenic and disrupt the natural hormone cycle is really concerning.[51]

Some of these chemicals break down the epidermal (skin) barrier. If the body is toxic, it doesn't have the energy stores to maintain the barrier.

[49.] This really beggars belief that a baby care product would contain an active ingredient harmful to babies.

[50.] Strenuously denied by manufacturers, it would seem possible that not just the ingredients but the act of blocking the pores in the underarm and genital areas might be a risk factor, albeit not a purely causative one.

[51.] If in any doubt check on the Environmental Working Group list at http://www.ewg.org/skindeep/. There is a comprehensive list there.

Increasingly we are seeing a breakdown in this barrier, with a lack of exposure to dirt, swimming outside, and so on, exacerbating it. You can get tested for any of these products and for the by-products of bacteria and yeast; organic acids, for instance, like D-arabinatol and D-lactate. There are some stool tests too, but as all testing is generally expensive, it is generally best to assume toxicity rather than not, as you cannot avoid these toxins in daily life.

Skin Solutions

Looking on the more positive side; what can we do about this? Well, first we can add organic natural yogurt to skin (or various dairy free versions if you prefer). Topical probiotics are a growing area of research. Bifidobacterium and Lactobacillus are the main ones for internal consumption and many probiotic supplements contain these. Using the more holistic approach to recommend foods rather than supplements, it would be better to use fermented foods containing them, for example, kefir.

Any skin condition has a gut bacterial imbalance (dysbiosis) associated with it, for example acne, psoriasis, and so on, are direct indicators that the commensal bacteria (the ones that live with us in harmony) are imbalanced. This can be due to an overgrowth of other, less beneficial strains but can be influenced by stress too. By adding bacterial strains such as Lactobacilus directly to skin that is, *topically*, they can colonize us more effectively. They then produce lactic acid that lowers the pH of skin (which is good, as it should be normally acidic). By improving the colonization of these commensals, they can then act as a barrier to less welcome pathogens. The range of flora depends on your age, race, location, and environment, but there is much you can do to improve it.

The Microbiome and the Brain

We have long ignored the importance of our microbiome in our general health and well-being, but particularly ignored is the importance of the gut microbiome in mental health where it has been assumed to have no connection whatsoever. However, we are finding out that this is simply not true. From the study of functional neurology, we are learning that there are vast correlations between the health of your gut and the health of your brain. This differs from traditional neurology as it looks at the fundamentals of health rather than just being based on symptom management.

One of the common factors in the health of both brain and body is a small carbohydrate called butyrate (produced by gut bacteria) which is the primary fuel for colonic cells and is also a powerful signaling molecule throughout the body. It is vitally important for maintenance of blood/brain barrier (BBB)—this is a built-in barrier to large molecules entering the *cerebrospinal fluid (CSF)*. If we compromise this, we increase leakiness of the brain. This could be the mechanism for Alzheimer's disease, for instance, and is maybe where autism and gut barrier function may be linked.

As we have already seen, we and our microbiome are interlinked; our food choices, our satisfaction, our mental health especially our mood, and so on are all governed by the health of our gut bacteria. We are now seeing the world through a lens filtered by the microbiome. If your butyrate levels are low, it is best to add butyrate not as a supplement but with probiotic bacteria that produce it. By eating more coconut oil, you naturally produce more as you encourage the bacteria that produce this to become more prolific. Even more controversially we can add directly via *fecal transplants* and enemas. Now, don't balk at this. These methods are becoming much more popular as they plainly work, despite our squeamishness. Inoculating ourselves with the bacterial colonies from a healthy person can reverse years of allergies and intolerances very quickly; I have witnessed this in someone I knew, and it was remarkable.

A researcher named Max Niewdorp in Copenhagen claimed he has even reversed diabetes with fecal transplants [Nieuwdorp14]. He concurs with the preceding statement by stating that changes in the microbiome to increase butyrate producers were the most potent method of doing this. This simple change reduces inflammation, as measured by the specific biomarkers of *C-reactive protein* (CRP) and *tumor necrosis factor* (TNF) in a cell signaling protein or cytokine—both clinical measures of inflammation. Moreover, it also improves the blood brain barrier function.

The brain is a wet organ full of fluids and therefore the drainage system needs to clear. There is a lymph system that does this in the brain from the CSF into the lymph vessels. CSF is part of a vast *extracellular matrix* (ECM) which takes up 20% of the space and can clog full of garbage if it is not kept clear. You know how bad things can get in your house if your drains are blocked? Waste products cannot be taken away and begin to putrefy. Well, it's not so different in your brain. There is another component to the ECM called *glial cells* or *microglia*; these help nerve cells to keep healthy—they are like caretakers in the school yard clearing out debris and making sure the brain cells (neurons) have enough nutrients to function well. The ECM (especially glial cells) and the microbiome are interrelated; there are similar levels of modulation to keep things running well such as levels of oxygenation, pH regulation, viscosity, and stickiness. We are now beginning to realize we are not sterile inside our guts; if the balance is wrong, opportunistic organisms take over. Could this be true of the brain also? This picture is still to fully emerge, but it is suspected so; the ecosystem of the brain is supported by the lymph fluid system (coined, helpfully "glymphatic" = glial cells +, lymph) circulation.

Interestingly, glial cells retract 60% of their volume at night (during sleep), so lymphatic drainage channels can open up and the brain can dump the trash of misfolded proteins that accumulate during the day. The glymphatic system and how it interacts with the lymph system of the rest of the body is shown - in Figure 1.5.

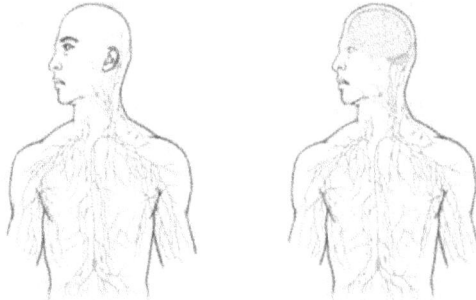

Figure 1.5. Glymphatic system—linkage with lymph.

If you are too acidic, you will have a harder time clearing stuff out as the brain's lymph circulation becomes sluggish. That is why good restful sleep is really important. Osteopathic or cranio-sacral (CS) therapy can greatly help this drainage. By doing this, we can slow down the progression of dementia, eliminate seizures, and restore the integrity of fluid flows at the structural level. But unfortunately, you won't see that offered within traditional health services. Some preventative/palliative work like this could do the work of the latest "'wonder drug" being developed for Alzheimer's or dementia for a fraction of the cost. Unfortunately though, the trials that have been conducted to support CS work (and shown great success), [Gerdner08] are small and largely unknown because there is no drug company money to support them. This is the story of modern evidence-based medicine; the evidence is very partisan.

There are other ways you can support your lymph drainage: exercise (which pumps the system, having no "heart" equivalent), yoga (with its emphasis on breath, stretches, and slow movement), an alkaline diet, and brain-friendly probiotics such as L. plantarum and Bifidobacterium species. There is even promise of vagal nerve stimulation (with electrical devices) and restorative breathwork (such as Wim Hof and others promote). This holistic, supportive approach is currently exploding as the options for drug interventions get less effective. I invite you to do your own explorations. Next, we are going to look at some specific cellular systems that get disrupted when things go wrong.

References

[Amin12] Amin, K. "The role of mast cells in allergic inflammation," *Respiratory Medicine* (2012): *106*(1) pp. 9–14.

[Appleton02] Appleton, N., *Rethinking Pasteur's Germ Theory: How to Maintain Your Optimal Health*. North Atlantic Books, 2002.

[Boonstra15] Boonstra E, de Kleijn R, Colzato LS, Alkemade A, Forstmann BU, Nieuwenhuis S. "Neurotransmitters as food supplements: the effects of GABA on brain and behavior". Frontiers in Psychology: (2015) 6(6) p.1520.

[Brizendine17] Brizendine, L. The Neuroscience Summit. Audio recording. Own notes, 2017.

[Bus15] Bus, James S. "Analysis of Moms Across America report suggesting bioaccumulation of glyphosate in U.S. mother's breast milk: Implausibility based on inconsistency with available body of glyphosate animal toxicokinetic, human biomonitoring, and physico-chemical data," *Regulatory Toxicology and Pharmacology* (2015): 73(3), pp. 758–764.

[Carnahan13] Carnahanm J. "MTHFR Gene Mutation-What's the Big Deal About Methylation?," Dr. Jill blog, May 2013.

https://www.jillcarnahan.com/2013/05/12/mthfr-gene-mutation-whats-the-big-deal-about-methylation/

[CSC00] "Carcinogens in Cosmetics." Campaign for Safe Cosmetics, 2023 https://www.safecosmetics.org/chemicals/known-carcinogens/

[Coghlan16] Coghlan, A. "Gut bacteria spotted eating brain chemicals for the first time," *New Scientist News,* June 2016. https://www.newscientist.com/article/2095769-gut-bacteria-spotted-eating-brain-chemicals-for-the-first-time/

[Cressey15] Cressey, D. "Widely used herbicide linked to cancer." *Nature News,* 2015.

[Daulatzai15] Daulatzai, M. A. "Non-celiac gluten sensitivity triggers gut dysbiosis, neuroinflammation, gut-brain axis dysfunction, and vulnerability for dementia," *CNS and Neurological Disorders Drug Targets* (2015): 14(1) pp. 110–131.

[Dinan 15] Dinan, T. G., Stilling, R. M., Stanton, C., and Cryan, J. F. "Collective unconscious: How gut microbes shape human behaviour," *Journal of Psychiatric Research* (2015): *63*, pp.1–9.[Domínguez-Bello08] Domínguez-Bello, M.G., Pérez, M.E., Bortolini, M. C., Salzano, F. M., Pericchi, L. R., Zambrano-Guzmán, O., and Linz, G. "Amerindian Helicobacter pylori strains go extinct, as European strains expand their host range", *PLoS ONE* (2008): *3*(10), e3307.

[Douillard17] Douillard, J. *"Eat Wheat: A Scientific and Clinically Proven Approach to Safely Bringing Wheat and Dairy Back into Your Diet."* Morgan James Publishing, 2017.

[Douillard20] "The Domino Effect of Ignoring Your Lymphatic System," 2020 https://lifespa.com/diet-detox/lymphatic-system-ayurveda/

[Ferrières04] Ferrières, J. "The French paradox: Lessons for other countries," *Heart* (2004): *90*(1), pp.107–111.

[Galland14a] Galland, L. The Gut Microbiome and the Brain, Foundation for Integrated Medicine, New York, 2014.

[Galland14b] Galland, L. "The gut microbiome and the brain," *Journal of Medicinal Food* (2014): *17*(12), pp. 1261–1272.

[Gerdner08] Gerdner, L., Hart, L. K., and Zimmerman, M. B. "Craniosacral still point technique: Exploring its effects in individuals with dementia," *Journal of Gerontological Nursing* (2008): *34*(3), pp. 36–45.

[Gillings15] Gillings, M. R., Paulsen, I. T., and Sasha, G. T. "Ecology and evolution of the human microbiota: Fire, farming, and antibiotics," *Genes* (2015): *6*(3) 841–857.

[Guyton15] Guyton, K. Z., Loomis, D., Grosse, Y., El Ghissassi F., Benbrahim-Tallaa, L., Guha, N., Scoccianti, C., et al. "Carcinogenicity of tetrachlorvinphos, parathion, malathion, diazinon, and glyphosate." *The Lancet Oncology:* (2015): *16*(5) pp. 490–491.

[Halvorsen10] Halvorsen, B. *Good Teeth for Life.* iUniverse, 2010.

[Hamilton01] Hamilton, G. "Dead man walking," *New Scientist,* 2303 (August 21): 30–33.

[HDC00] "Hormones Factors in Fetus Gender and Child Future Development." Health Doctrine. http://healthdoctrine.com/hormones-factors-in-fetus-gender-and-child-future-development/

[Honeycutt14] Honeycutt, Z., and Rowlands, H. (2014). "Glyphosate testing full report: Findings in American mothers' breast milk", Urine and Water. Moms

Across America and Sustainable Pulse, April 7, 2014. https://www.MomsAcrossAmerica.com

[Huddleston14] Huddleston, J. R. "Horizontal gene transfer in the human gastrointestinal tract: Potential spread of antibiotic resistance genes. *Infection and Drug Resistance* (2014): *2014*(7): pp. 167–176.

[Hutter15] Hutter, T., Gimbert, C., Bouchard, F., and LaPoint, F.-J., Being human is a gut feeling," Microbiome (2015): 3(9).

[Jan16] Jan, Y. H. "Novel approaches to mitigating parathion toxicity: Targeting cytochrome P450-mediated metabolism with menadione," *Annals of the New York Academy of Sciences. (Aug 2016): 1378*(1), pp. 80–86.

[Lane14] Lane, N. "Bioenergetic constraints on the evolution of complex life," *Cold Spring Harbor Perspectives on Biology (May 2014): 6*(5), a015982.

[Levine10] Levine, P. *In an Unspoken Voice, How the Body Releases Trauma and Restores Goodness.* North Atlantic Books, 2010.

[Lindsay23] Lindsay, S. Smith, S; Yang, S; Yoo, J. "Community Water Fluoridation and Rate of Pediatric Fractures." Journal of the Academy of American Orthopedic Surgeons: Global Research and Reviews (2023): 7(10), e22.00221.

[Maciel15] Maciel, T. T., Moura, I. C., and Hermine, O. "The role of mast cells in cancers," F1000 *Prime Reports* (2015): 7(9).

[McFall-Ngai13] McFall-Ngai M. et al., "Animals in a bacterial world, a new imperative for the life sciences." *Proceedings of the National Academy of Sciences.* (2013): *110*(9), pp. 3229–3236.

[Mercurio14] Mercurio, P., Flores, F., Mueller, J. F., Carter, S., and Negri, A. P. "Glyphosate persistence in seawater," *Marine Pollution Bulletin* (2014): 85(2), pp. 385–390.

[Montiel-Castro13] Montiel-Castro A. J., Gonzalez-Cervantes, R. M., Bravo-Ruiseco, G., and Pacheco-Lopez, G. "The microbiota-gut-brain axis: Neurobehavioral correlates, health and sociality," *Frontiers in Integrative Neuroscience* (2013): 7(70). https://www.frontiersin.org/articles/10.3389/fnint.2013.00070/full

[Myers16] Myers, J. P., Antoniou, M. N., Blumberg, B., Carroll, L., Colborn, T., Everett, L. G., Hansen, M., et al. Concerns over use of glyphosate-based herbicides and risks associated with exposures: a consensus statement. *Environmental Health:* (2016): 15(19).

[Nagel10] Nagel, R. *Cure Tooth Decay: Heal and Prevent Cavities with Nutrition.* Golden Child Publishing, 2010.

[Naviaux16] Naviaux, R. "Mitochondria and Autism," July 15, 2016, YouTube video. https://www.youtube.com/watch?v=Z5NFKWli8yE&t=3442s

[Nelson 16] Nelson, A. "Glyphosate unlikely to pose Risk to Humans UN WHO study says," The Guardian, May 16, 2016. www.theguardian.com/environment/2016/may/16/glyphosate-unlikely-to-pose-risk-to-humans-unwho-study-says

[Nieuwdorp14] Nieuwdorp, M. "Fecal transplantation in obesity/type 2 diabetes." *Endocrine Abstracts,* (2014): 35: S12.2.

[Pert99] Pert, C.. *Molecules of Emotion.* Simon & Schuster, 1999.

[Pollan09] Pollan, M. *In Defence of Food: The Myth of Nutrition and the Pleasures of Eating: An Eater's Manifesto.* Penguin, 2009.

[Pollan13] Pollan, M. *"Cooked: A Natural History of Transformation.".* Penguin, 2013.

[Price09] Price, W. *"Nutrition and physical degeneration.* Price Pottenger Nutrition, 2009.

[Sagan67] Sagan, L. "On the origin of mitosing cells," *Journal of Theoretical Biology (1967): 14*(3), pp. 255–274.

[Samsel13] Samsel, A., and Seneff, S. "Glyphosate's suppression of cytochrome P450 enzymes and amino acid biosynthesis by the gut microbiome," *Pathways to Modern Diseases Entropy (2013): 15*(4), pp. 1416–1463.

[SIS15] "A Roundup of Roundup® Reveals Converging Pattern of Toxicity from Farm to Clinic to Laboratory Studies." Science in Society (January 19, 2015). http://www.i-sis.org.uk/Roundup_of_Roundup.php

[Stilling 14] Stilling, R. M., Bordenstein, S. R., Tinan, T. G., and Cryan, J. F. "Friends with social benefits: Host-microbe interactions as a driver of brain evolution and development?," *Frontiers Cellular and Infection Microbiology* (2014): *4*(Oct. 27): p. 147.

[Theoharides13] Theoharides T. C., Asadi, S., Panagiotidou, S., and Weng, Z. "The 'missing link' in autoimmunity and autism: Extracellular mitochondrial components secreted from activated live mast cells." *Autoimmunity Reviews* (Oct 2013): *12*(12), pp. 1136–1142.

[Vanuytsel13] Vanuytsel, T., Vermier, S., and Cleynen, I. "The role of Haptoglobin and its related protein, Zonulin in inflammatory bowel disease," *Tissue Barriers* (2013): *1* (5) e27321.

[Vedin14] Vedin, O., Hagström E., Gallup D., Neely M. L., Stewart R., Koenig W., Budaj A., et al., "Periodontal disease in patients with chronic coronary heart disease: Prevalence and association with cardiovascular risk factors," *European Journal of Preventive Cardiology* (2014): *22*(6) pp. 771–778.

[Wahls16] Wahls, T. "Kellman Center blog." https://www.kellmancenter.com/post/an-exclusive-interview-with-dr-terry-wahls /. January 2016.

[Wang00] Wang W., Uzzau, S., Goldblum, S. E., and Fasano, A. "Human Zonulin, a potential modulator of intestinal tight junctions, *Journal of Cell Science.* (2000): *113*(24), 4435–4440.

[Watts05] Watts T., Berti, I., Sapone, A., Gerarduzzi, T., Not, T., Zielke, R., and Fasano, A. Role of the intestinal tight junction modulator zonulin in the pathogenesis of type I diabetes in BB diabetic-prone rats, *Proceedings of the National Academy of Sciences (2005): 102*(8), pp. 2916–2921.

[Wikipedia1] "Intrinsic factor." Wikipedia. Last modified November 7, 2023. https://en.wikipedia.org/wiki/Intrinsic_factor

[Wikipedia 2] "Dendritic cell." Wikipedia. Last modified on October 10, 2023. https://en.wikipedia.org/wiki/Dendritic_cell

[Willacy23] "Listeria." Patient. http://patient.info/health/listeria-leaflet

[WHO03] WHO. "Social determinants of health," World Health Organization, 2003. https://intranet.euro.who.int/__data/assets/pdf_file/0005/98438/e81384.pdf, *2003.*

Yong, Ed. *I Contain Multitudes: The Microbes Within Us and a Grander View of Life.* Bodley Head, 2016.

[Zick10] Zick C.D., and Stevens, R. B. "Trends in Americans' food-related time use: 1975–2006." *Public Health Nutrition* (2010): *13*(7), pp. 1064–1072.

CHAPTER 2

THE REGULATION OF HEALTH

Now that we have looked at the diversity of our microbiome in all its magnificent glory and understood it is an organ system just as important as our heart, liver, or brain, we can begin to see that health is more than just getting our good bacteria in place. Health is a dynamic, synergistic relationship between the microbial universe in our guts, and our cells' metabolic performance. We now turn to that metabolic link with the microbiome.

Mitochondria and the Microbiome

We are energy requiring beings; unlike plants we cannot make energy directly from the sun via chlorophyll. We have a different system using mitochondria that we inherited by using bacteria to give us an efficient energy production "engine." This is a really interesting story of symbiosis that happened many millions of years ago (as described in Chapter 1).

Mitochondria might not be well known to you, unless you've studied biology—even then, you've perhaps not understood them in the context of health. They are tiny organelles that exist in every cell of the body (except red blood cells; see Figure 2.1). They are sometimes called the "powerhouse" of the cell, as they produce energy, but in fact they are far more than that. However, in order to discuss their relevance, we are going to have to start with that basic function as it is key to everything else that follows.

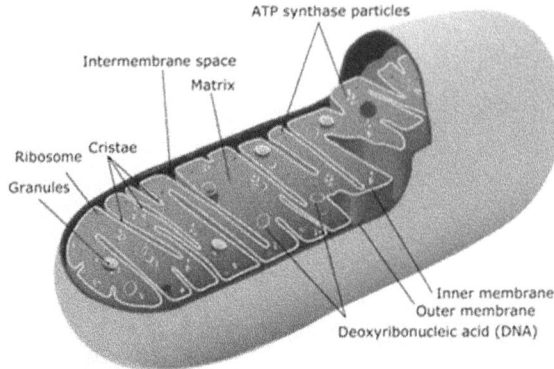

Figure 2.1. The mitochondrion.

Mitochondrial Energy Production

Energy production is essential for every process that happens in the body; the mitochondria is where it happens. Each cell has between 100 and 100,000 mitochondria (except red blood cells as already mentioned). We are only just discovering how important they are to our functioning.

Figure 2.1 shows you the shape and structure of a typical mitochondrion. Note the lozenge shape and convoluted interior membranes which increase the surface area for reactions to take place.

Mitochondria produce fuel by combining certain nutrients with a cascading chain of protein complexes which act as electron donors and receivers. This is the *electron transport chain* (ETC) as illustrated in Figure 2.2.

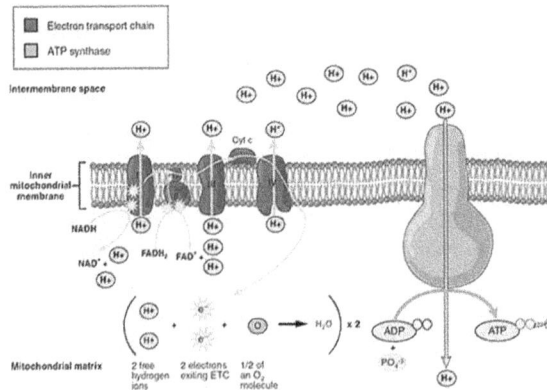

Figure 2.2. The electron transport chain

This has been formulated in the form of an "engine" that pumps electrons and protons through a series of trans-membrane protein complexes which ends in the production of *adenosine triphosphate* (ATP), the energy molecule. This process produces dangerous free radicals as part of the process (elements with an unpaired electron), likened to overexcited particles—an excess of which cause cellular damage. It is therefore good to avoid adding energy into the system just before you go to bed, in other words don't eat before bed! Leave at least three hours before bedtime to have your last meal. This may explain why restricting calories and some gap between (intermittent fasting) has been shown to have beneficial effects. Time your sugar /carbohydrate intake before exercise not before bed.

However, that is only part of the story; the mechanistic "pump" view has already shown to be outdated, part of the legacy of Victorian science. In fact, it is not protons that do the work, but some element of photons (light particles) interacting with super accelerated electrons. This harnesses "quantum coherent energy transfer between uniquely arranged chromophores in light harvesting photosynthetic complexes" [Craddock14, para. 1]. In other words, this energy production is a *quantum process* not a mechanistic one. This is one of the new understandings we are reaching about quantum biology (discussed in more detail in Chapter 7 later). It is an exciting development.

Mitochondrial Intracellular Communication

According to the latest thinking, mitochondria are not just powerhouses, but influence vital bodily functions such as memory and aging through to combating stress and disease [Hamilton14].

In addition to its well-recognized role as the "engine" of the cell, recent studies have indicated that the mitochondria control many more cellular functions. For example, the mitochondrial network in the cell appears to be electrically coupled and coordinates such critical processes as cell death (apoptosis), cell proliferation, autophagy (self-destruction), cell differentiation, and aging. The view seems to be emerging that perhaps the mitochondrial network is actually the "brains" of the cell, controlling much of the cell's processes, including activities of the nucleus [McFall-Ngai13].

Mitochondria have their own DNA (or "'genome") inherited from the maternal line only. This makes them unique and very different to the nuclear DNA which is a combination of your mothers and your fathers. Mitochondrial (mt) DNA can talk to your nuclear genome, and both can communicate to the DNA in other cells. This communication system gets fouled by toxicity both within and outside the cells. For example, with an inflamed cell membrane, hormones can't dock onto receptors and can't deliver the message into the cell or get into the cell itself. Cells become inflamed and toxic causing inflammatory molecules called cytokines to be released. At a low lower level of inflammation, the membranes of the mitochondrion become inflamed, and the cell can't make energy—ATP production is affected. If that goes on long enough, the immune system which has priority for your energy, gets mixed up and can't identify friend from foe and *auto-immune disease* (AID) results.

There are other complications from specific hormones which can't be "read" properly, for example, the appetite regulating hormone leptin is also blocked—the hypothalamus in the brain which reads these signals can't recognize when you need to stop eating. So, not only do you eat more than you need, but your fat cells get bigger too with the inevitable result that you gain weight. Rather than pumping the body with excess hormone, we need to fix the cell membrane which is the root cause of the dysfunction.

Chronic fatigue syndrome is another very common chronic illness that results from these dysfunctional messages: such is the free radical damage in the mitochondria that the body can't step up its energy production when needed. This causes the primary symptom of fatigue that does not get better with rest (as the ATP can't recycle properly); joint aches also result from this primary issue. Cancer induction may also stem from this disrupted mitochondrial signalling and cell division control—this is discussed later in this chapter when we read about quantum effects.

Stress is a major component to dysfunctional mitochondrial functioning. The true meaning of the word is stress, coined first by Hans Selye, is something that is outside of normal homeostatic control. Homeostasis is the ability of the cell to maintain an internal balance. So, it is anything that shifts the body outside of its normal range of physiological balance. Scientists distinguish between "good stress" (eustress) and "bad stress" (distress). In fact, the word stress is often bandied about as exclusively bad, but light stress has a positive effect on homeostasis. However, if a certain point is passed then the systems that respond to the environment, especially those of mitochondrial signalling, start to get overwhelmed. As quoted by Alistair Nunn and others: Overall, As researcher Alistair Nunn and others have found, it seems that mitochondrial fusion induced by mild stress or reduced nutrients tends to enhance oxidative phosphorylation (the form of efficient energy production), whereas too much stress, excess nutrients, disease and inflammation, including cancer, induces fragmentation which usually leads to mitophagy (destruction) and reduced oxidative phosphorylation [Nunn2016].

The main point here is that the health of the mitochondria—and thus of our cells—is directly related to the gut microbiota and its control of inflammation. People with imbalance (dysbiosis) also develop problems with the membrane integrity of their mitochondria causing the triggering of an intracellular environmental switch called the "inflammasome" [Bauernfeind13]. Dr. Michael Ash has identified several types of microbial (bacterial, viral, fungal, and parasitic) "routes of influence." He classifies them as:

- commensal (the good guys)

- pathobiontic (the teenagers of our gut; they live with us, but you wish they didn't!)

- pathogenic (the bad guys)

His research has shown important differences in the gut diversity of Western people compared to native hunter gatherers of Tanzania (the Hadza people) who have a much greater variety of species in their guts compared to ours. This is undoubtedly due to the difference in lifestyle; they live with animals (mainly cows) and cover themselves with mud to protect against the sun. They also eat a massive 120 grams of fiber a day (compared to the average 9–10 grams in the western diet). In other words, they have massive exposure to animal and plant microorganisms. This makes an incredible difference in the types of species that inhabit their guts compared with ours. We have lost 50% of the bacterial species in our gut already compared to these tribal people. And microbial diversity, as we are beginning to see, gives us more protection against disease. It is no wonder that we are suffering more and more inflammatory disease as a result.

Genes Are Not Your Destiny: Epigenetics

There is another layer to the story of genetic regulation that has been coined *epigenetics*—it means "above genetics." This is the next layer of complexity as it helps determine which genes are expressed. This is a complete paradigm shift in our understanding of human complexity, and it goes some way to explaining why it is we have so few genes, compared to prior predictions based on that complexity compared to other species.

Remember, we have barely any more genes than a worm (28,000 compared to 24,000). That is certainly not enough to explain how we came to be so much more complex. Clearly there is something missing in this "genetic determinism" understanding. Indeed, the discovery of epigenetics shows us that it is not just the genes but *how they are read* that determines how well we live and age. In short, this means certain "environmental factors" turn the genes on and off in a complex orchestration between groups of genes. Much like the words in a sentence can be combined in a multitude of ways, the genes plus environment create different expressions.

So, not only do we have the massive contribution of gut microbe DNA (and all the other microbes on your skin and mouth, etc.) but we also have mechanisms which change our ability to adapt or compensate to subtle changes in the environment. These mechanisms range from changes in the readability of the DNA by adding methyl groups to certain sections of the code,[1] to the way the DNA is wrapped and unwrapped via proteins called histones. I liken it to adding highlights to the blueprint: "ignore this bit" or "use this." As Cath Ennis says in her excellent summary piece on epigenetics in *The Guardian* newspaper:

> Even though every cell in your body starts off with the same DNA sequence ... the text has different patterns of highlighting in different types of cells—a liver cell doesn't need to follow the same parts of the instruction manual as a brain cell. But the really interesting thing about epigenetics is that the marks aren't fixed in the same way the DNA sequence is—some of them can change throughout your lifetime, and in response to outside influences. Some can even be inherited, just like some highlighting still shows up when text is photocopied. [Ennis14, paragraph 6]

The important point about these environmental epigenetic regulators is that they can *adapt much more quickly than genetic changes* are able to. Hence if the body needs to shut down energy quickly to save itself, it is able to do so without waiting for the inevitable time lag it takes to read the DNA and create new proteins. Since we know that some of these epigenetic changes are heritable[2] it begins to seem a plausible explanation for why some trauma

---

[1] In fact, methyl groups attach to the cytosine ('C') base of the base-pair alphabet—the "rungs" of the double helix ladder. This affects its readability by interactions with the histone protein coating that normally protects it.

[2] Most of the experimental evidence is in mice rather than humans; but it has been observed in the children and grandchildren of Holocaust survivors and those subjected to the Dutch "Hunger Winter" during WWII, for instance.

passes down the generations in the form of susceptibility to stress and long-term health problems.

Our environment has changed rapidly in the last sixty years or so; more rapidly than at any point in our history. There are more toxins, stress, poor quality food, disconnection with the earth and its seasons, and so on, that mean we are getting discordant messages which the body struggles to cope with. Of all the messages, that which we take directly into our bodies, food, is perhaps the biggest epigenetic modifier of all (although stress comes a close second and may supplant it if chronic).

The human genome and the epigenetic expression of that genome are based on what it learned thousands of years ago from whole, natural foods. Eventually science will prove that refined sugar and partially hydrogenated vegetable oils are the leading health destroyers of our times as both are highly inflammatory for cell membranes - both outer cell membrane and inner mitochondrial membranes. Unfortunately, these two substances are primary ingredients in our diets today adding cheap bulk to nutritionally depleted foods and, because of poorly regulated industrial food production are allowed while cholesterol is denigrated[3] [Tips20].

The Epigenome

Both our native cells and those of the microbiome are regulated by the *epigenome* (the multitude of chemicals that regulate the genome); they are one unit. Genetic expression can be influenced by nutrients, toxins, and even thoughts and feelings that trigger the stress response and immune systems. When they are overtly negative, they can become fixed in a habitual pattern of threat to the system which can even be transmitted to the next generation. Even the way we are conceived seems to have epigenetic consequences; a process called epigenetic imprinting adjusts the activity of genes that will shape the character of the child yet to be conceived [Lipton05]. Whether this is solely by epigenetics or by behavioral influences is a moot point; it probably involves both. What is undisputed is that we can influence our children and their children by the way we lead our lives and take care of our relationship with ourselves.

[3.] Cholesterol is needed by the body for hormone and Vitamin D production and by the nervous system for its cell membranes. Cholesterol test results actually measure the lipoprotein complex that the body forms to transport it around. It is excess sugar and poor-quality fat that promotes inflammation causing high cholesterol not a high-fat diet.

The implications of this realization for our interaction with our microbiome are *huge*. If 2.2 million genes come from the microbiome compared to the 23,000 million from our own DNA, then *99% of the DNA in our bodies is microbial*! Just think of the multitude of epigenetic fine tuning that is possible with this amount of extra DNA and protein interactions. Without these microbes, you would be unable to speak a sentence, or think. This intimate interaction regulates our functioning more than our own genes, so it is important to get the balance of flora right and respect their input into our evolutionary history. Every cell in your body is the entire history of evolution in those genes. As Deepak Chopra said, "Your ancestors are alive in you!" Once we realize, we are constantly passing on epigenetic modulations onto the activity of genes we had better take care of what environments we subject our cells to. Let's now have a look at how cells determine their environment and interact with it.

The Cell Membrane: Sensor of the Cell

In the ground-breaking book *The Biology of Belief* by Dr. Bruce Lipton, this cell biologist turned author and campaigner for saner medicine, talks about the "ah-ha moment" when he worked out that it wasn't the nucleus (containing the chromosomal DNA) of the cell that controlled function, but the cell membrane. This may be news to any of you who, like me, studied biology at school. Trust me, despite not having filtered down yet to school teaching, it is actually a very well-supported argument.

He had been studying the properties of the phospholipid bi-layer and the integral proteins that straddle the lipid "sandwich" and realized that the nature of the polar heads created an organic liquid crystal and the proteins allowed communication of information from inside to outside. This exactly mimics the silicon chip in your computer. He believed he had found the secret of why a cell, when deprived of its nucleus, will happily continue living until it runs out of proteins - but if the cell membrane is blocked, it dies pretty quickly. Without information flow, the cell is no longer tuned to its environment and cannot survive.

Furthermore, he began to study the nature of quantum physics and realized that it "is relevant to biology and that biologists are committing a glaring, scientific error by ignoring its laws" [Ibid]. Much current medicine has relied on an outmoded Newtonian approach to the biochemistry of the cell: molecules colliding into one another like billiard balls to create chemical reactions. This is a view based on a theory of physics that is at least

200 years old. Although thankfully, that is changing, with biological systems now being studied through the lens of complexity theory and the idea of emergent properties of complex organized systems. Consciousness is now being viewed this way, for instance as an emergent property of information flow. I go into the importance of this new way of thinking for the healing power of belief and understanding of the nature of consciousness in the final chapter, but here I want to deal with how it relates to the workings of the cell and therefore metabolism, that is, "where quantum physics meets biology" [Arndt09]. This is an exciting new area of research which has many implications for our microbial world within.

Quantum Cellular Processes

When we apply a Newtonian model[4] to the cell (where everything it reducible to its mechanistic interactions) we look at pathways of interaction between the cell's protein signaling molecules as a linear chain of information flow A - B - C and so on. But if, as now seems more likely, we consider the flow of information as *a complex web of dynamically interacting energies* within the intracellular space (with water as the matrix), we find that perturbation of any one part of that web causes ramifications throughout all of the interconnecting parts, much like ripples on a pond. This is a *holographic* view of the cell where all the parts contain the whole, more closely matching how living cells and bodies really work.

This might explain why introducing a drug specifically designed to alter one aspect of the pathway has such large, unintended effects throughout the body (and why pharmaceuticals have such long lists of side-effects). It just isn't based on an accurate understanding of how the cell works to add such grossly altered chemicals (they can't be natural to be patented so must be altered in some way to become licensed as a drug). These slightly altered molecules are similar enough in structure to the parent (naturally occurring) molecule but their energy vibration is different so the body has trouble dealing with them in the same way as natural molecules. This is a subtle but significant difference.

Modern medicine has not only tied itself inexorably to the mast of pharmaceutical intervention as the first line attack on disease, but it has

---

[4.] Issac Newton, pioneering physicist who viewed all processes in the universe as being calculable and reducible to simple forces of attraction and repulsion involving gravity and mass. He was limited in his understanding by the technology of measurement at the time (advances in which later disproved his theories along with the radical new theories proposed by Einstein).

studiously ignored and derided alternatives such as *energy medicine* which deal with these subtler but more harmonious methods. The only scientists who could support the basis of energy medicine (physicists) are generally not that interested in the subject and don't interact with the medical establishment in any case. So, we are left with the "physicalist" approaches of biochemists, cell biologists, and neuroscientists, that lead to yet more research of "magic bullets" and "wonder drugs" or, more recently, gene therapy (the latest manifestation). That is not to say that some drugs don't have their intended effects, but they do so at a cost to the rest of the body as they affect so many systems. The body can adapt and attempt to reduce the damage, as it is intelligent, but there is a limit to how much it can ameliorate the effects of these very blunt instruments. Furthermore, the overuse of prescription drugs to silence our symptoms is like turning off the warning light of the car dashboard—it takes away the immediate effect but does nothing to address the cause. Even more concerning is the idea that the reliance on them takes away our personal responsibility for our health.

I dare to raise the issue of the homeopathic approach here which has borne the brunt of ridicule by mainstream medicine. The main bone of contention has always been that homeopathic solutions are diluted to such an extent that hardly any molecules of the original substance remain in the water. But when we understand that water holds the energy signature of the molecules that have once been dissolved in it, we gain a potential mechanism for the harmonizing effect of such remedies. I should say I am not a homeopath myself, but I know many people who have used it to good effect, and I see no reason to denigrate something that we do not fully understand unless it does active harm (which you could say about pharmaceuticals anyhow). To say, as many critics do, that it is just placebo effect (whereby belief alone makes it work) is not a full explanation as it works with animals too. The absolute purpose of the scientific method is, after all, to find an adequate explanation or theory that comes the closest to explaining our observations. I believe that we may now be close to finding that theory as we evolve our understanding of the properties of water.

The Fourth Phase of Water

Recent exciting research by Gerald Pollock's team in the United States has also highlighted that biological water is different to water found elsewhere; it forms a fourth phase (separate from gas, liquid, or solid phases) called the *gel phase*, which forms an *exclusion zone* (EZ): a negatively charged interior layer surrounded by protons. EZ water acts like a battery and, with infra-red energy (from metabolic processes and daylight), helps to propel biological processes like blood flow. This explains the paradox that some researchers have found; how a heart can pump blood all around the body given the size of the pressure wave. It just isn't possible—calculations by biophysicists have shown it would have to be at least one hundred times stronger! There is some force missing from the equations and Pollock thinks this may be provided by EZ water pulling blood through the capillaries. This biological water may also hold memory as the inevitable quantum effects of a polarized entangled system that holds information. This is true, also, of DNA, Vitamin D, and all the chromophores (natural color giving molecules) present in the body that harvest light.

Electromagnetic fields (EMFs) generated by mitochondria could generate such "water order," and thus protect against de-coherence and diseases such as cancer [Pokorny13]. Far from being too warm and wet for quantum effects, it seems that living systems have optimized them by some means as yet unknown but seemingly related to this fourth phase of water.[5] Quantum computers seem such an amazing breakthrough but biological life has been using them since the beginning so we have a bit of catching up to do.

Microtubules as Quantum Sensors

Some theorists have attributed a quantum nature to parts of the cell called *microtubules* (small hollow tubular strands of protein filaments involved in structural support and transport of materials within the cell cytoplasm and division of cells).

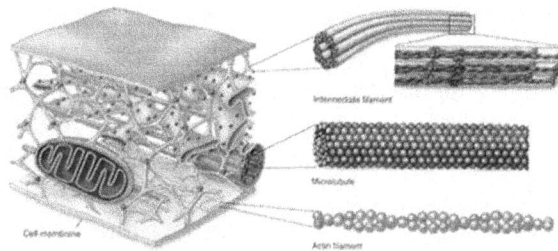

Figure 2.3. Microtubules.

During cell division, they are intimately involved in pulling apart the two arms of chromosomes of our nuclear DNA. They form the structural skeleton of the cell and "platforms for intracellular transport" through their hollow tubes. However, there may be more intelligent processes at work.

[5.] For further discussion, please see *Life on the Edge* by Jim Al-Khaleli.

Quantum biologists Roger Penrose and Stuart Hameroff, who have been at the forefront of this work, have postulated:

> The "tubulin" subunit proteins, which comprise microtubules, also possess a distinct architecture of chromophores, namely aromatic amino acids, including tryptophan. The geometry and dipolar properties of these aromatics are similar to those found in photosynthetic units indicating that tubulin may support coherent energy transfer. Tubulin aggregated into microtubule geometric lattices may support such energy transfer, which could be important for biological signalling and communication essential to living processes." [Craddock14, Abstract][7]

The Microbiome and Quantum Entanglement

Since we are now finding quantum molecular processes in photoactive (light harvesting) biological systems like Vitamin D, chlorophyll, as well as the magnetic field orientation of birds (used in migration), DNA mutation, neuronal signaling and vision to name but a few [Tamulis14], it comes as no surprise that our microbiota uses them too. It seems viruses detect a suitable host by "sensing" the quantum oscillations in the vicinity, using their protein coat as a sort of antenna (much like chlorophyll does) [Hameroff,16]. In fact, the origin of the microtubules themselves may be from a spirochete (ciliated organism) that the first eukaryotes (animal cells) absorbed much like those that formed mitochondria. This area of quantum biology is about to explode our ideas of what constitutes life and consciousness [Thompson10]. This is something we will consider in more detail in Chapter 7.

Interestingly, it is now becoming clear that bacteria can transfer electrons both between the same species and with other species in a form of symbiosis via "bacterial nanowires." In effect, these are biological conductors; they can transfer energy [Nunn2016] both through the more classical idea of electron transfer and via a more quantum style effect such as tunneling that describes how objects can "jump" energy barriers without the necessary energy because they can exist as probability waves not particles; a form of wave coherence. It has been observed in the way electrons travel along enzymes for instance and is almost certainly how the electrons in the ETC of the mitochondrion generate energy. These effects "seem to be enhanced by particular molecules, such as the aromatics (molecules with alternating high energy double bonds forming an electron cloud (...) which are key components of respiratory chains" [Nunn2016, Introduction paragraph 13].

This brings me to a basic proposition of quantum physics that may astound you—quantum particles like electrons and photons show what is called "superposition"; they can be either a wave or a particle depending on whether they are observed or not.[6] This idea breaks down in the regular, macroscopic world due to interactions with the random molecules of the environment. Hence a cat, far from being likely to disappear, as in Schrodinger's famous theorem,[7] is in fact intimately coupled to being in its environment. When we observe the cat, there is a "collapse of the wave function" into matter; it appears solid and real (which it is for our purposes). I will not go into this in more detail, but for a basic introduction to quantum concepts briefly discussed here, see *30-Second Quantum Theory* [Ball14]. To say that these ideas are revolutionary is an understatement; I can barely get my head around them so if you are interested in them, I would encourage you to read more.

I am more than ever convinced that we are only seeing what our logical brains have been taught to understand, leaving aside our much more powerful intuitive capacity (perhaps a more subconscious facility). Rudolf Steiner, (founder of the Steiner school's movement and far more), understood this and his approach to health and healing in the early part of the last century was far in advance of his time. But he was not alone—there were many doctors, philosophers, and writers of that period who understood that nutrition was far broader than the food one eats and takes in many aspects of a person's spiritual and physical life [Fergusson08]. Sadly, this was obliterated by the rise in industrial medicine of the 1950s and 1960s, but we are finally seeing a resurgence, I think, of the understanding that our cells are more than just a machine.

That is not to say that a Newtonian model is entirely without merit, it is just that it tends to apply at larger (macro) levels rather than at the scale of very small things like cells and molecules. However, one area where medical science has embraced the quantum model is in the area of scanning technology and we see great advances in the use of electromagnetic frequencies in laser surgery and fMRI scanning, for instance. This uses the wave nature of EMFs to interact with the natural energy signature of the body. But for some reason this understanding has not extended to the

---

[6.] Observation here means not just a human observer but a machine that a human operates to observe. So, for instance, a Geiger counter would be one such device.

[7.] The physicist Erwin Schrodinger postulated that, according to quantum theory, a cat in a box can be in two states simultaneously, that is, either dead or alive and only "fixes" itself toward one of the other state when the box is opened and observed.

use of natural body energies or systems of detection. For instance, it is my belief that Reiki and other energy techniques use the ability of our senses to detect and manipulate energies of other living systems. We are at the beginning of our understanding of energy medicine, but it is an expanding field.

The Brain-Gut-Microbiota Axis in Mental Health

For years we have known that the gut has a neurological network termed the *gut-brain* or sometimes *enteric brain* that possesses vast neuronal connections (100 billion neurons—more than in the spinal cord). However, we now have to extend this model to include the influence of the microbiota—hence the new term *brain-gut-microbiota axis* (BGM) [Zhou15]. This involves elements of the central nervous system (CNS), the neuroendocrine and neuroimmune systems, the sympathetic and parasympathetic arms of the autonomic nervous system (ANS), the enteric nervous system (gut nervous system) and, of course, the microbiome.

This new understanding recognizes that *neurological function is affected by the microbiome*—a departure from most current teaching in medical schools and how we treat mental disease. The gut and its bacteria are inextricably and intricately interwoven with brain function. This flies in the face of all we have previously been taught that is, *that a problem in the brain has its root in the brain*, for example, as in the study of Alzheimer's disease which has been almost exclusively looking at problems in *brain* physiology. But this approach is not making very much progress in controlling that disease, in fact it could be said we are losing ground.

When we understand that this integrated system is highly influenced by *the body's* chemistry, we can look to what controls this; primarily our nutrition and lifestyle that either promotes or reduces inflammation in the body.[8] So, we need to address what David Perlmutter, expert in brain physiology, calls, "root-cause resolution practices." In other words, we need to look beyond the organ that is having the issue and turn to the systemic (whole body) up-regulation of inflammation. Dietary practice is the main influence which triggers inflammatory cascades throughout body and brain. It is an old adage from Hippocrates, the father of modern medicine, that the "genesis of most disease is in the gut." The permeability of the gut lining is a cardinal player in this cascade as we have already seen. There

---

[8.] Although a good inflammatory response is part of the body's attempt to destroy pathogens and thus essential for survival, it can rapidly accelerate the aging process if it becomes chronic.

are several factors that increase gut permeability. Gluten is one of the main players (via the zonulin mechanism which opens up the junctions between cells). However, even without permeability, just gut dysbiosis (imbalance) can still give us inflammation.

It has been noticed that there are seasonal differences in inflammatory markers in the blood and cerebro-spinal fluid (CSF); with more proinflammatory cytokines present in the winter—up to 30% more! This has profound implications for psychiatric health and may explain some of the reasons behind Seasonal Affective Disorder (SAD) [Dopico15]. Inflammation is at the heart of most disease—but what causes the body to react in this way? In the next chapter, we will look at other threats to health that impact the microbiome.

References

[Arndt09] Arndt M., Juffman, T., and Vedral, V. "Quantum physics meets biology," *Human Frontiers Science Program Journal* (2009): 3(6), pp. 386–400.

[Ball14] Ball, P. *30-Second Quantum Theory: The 50 Most Important Thought-Provoking Quantum Concepts, Each Explained in Half a Minute.* ICON Books Ltd., 2014.

[Bauernfeind13] Bauernfeind F., and Hornung, V. "Of inflammasomes and pathogens—sensing of microbes by the inflammasome." *EMBO Moleculur Medicine,* (June 2013): 5(6) pp. 814–826.

[Craddock14] Craddock, T. J. A., Friesen, D., Mane, J., Hameroff, S., and Tuszynski, J. A. "The feasibility of coherent energy transfer in microtubules," *Journal of the Royal Society Interface,* (2014): 11(100), 20140677.

[Dopico15] Dopico, X. C., et al., "Widespread seasonal gene expression reveals annual differences in human immunity and physiology," *Nature Communications,* (2015): 6(1), p.7000.

[Ennis14] Ennis, C. "Epigenetics 101: A beginner's guide to explaining everything." *The Guardian* (April 25, 2014). https://www.theguardian.com/science/occams-corner/2014/apr/25/epigenetics-beginners-guide-to-everything

[Fergusson08] Fergusson, J. *The Vitamin Murders.* Portobello Books, 2008.

[Hameroff,16] Hameroff, S. *Ultimate Computing: Biomolecular Consciousness and NanoTechnology.* North Holland, 2016.

[Hamilton14] Hamilton, G. "The micromanagers." *New Scientist* (2014): 223(2987), pp. 42–45.

[Lipton05] Lipton, Bruce., *The Biology of Belief: Unleashing the Power of Consciousness, Matter and Miracles*, Cygnus Books, 2005.

[McFall-Ngai13] McFall-Ngai M. et al., "Animals in a bacterial world, a new imperative for the life sciences." Proceedings of the National Academy of Sciences. (2013): 110(9), pp. 3229–3236.

[Nunn16] Nunn, A., Guy, G. W., and Bell, J. D. "The quantum mitochondrion and optimal health," *Biochemical Society Transactions* (2016): 44(4) pp. 1101–1110.

[Pokorny13] Pokorny J., Pokorny, J., and Kobilková. "Postulates on electromagnetic activity in biological systems and cancer," *Integrated Biology* (2013): 5, pp. 1439–1446.

[Tamulis14] Tamulis, A., and Grigalavicius, M. "Quantum entanglement in photoactive prebiotic systems," *Systems and Synthetic Biology*, (2014): 8(2), pp. 117–140.

[Thompson10] Thompson, E. *Mind in Life*. Harvard University Press, 2010.

[Tips20] Tips, Jack. *The Cholesterol Myth: A Deception of Mammoth Proportions*. OpenBookHealth.com, 2018. https://wellnesswiz.com/wp-content/uploads/2018/11/OBH-Cholesterol-Myth-v1.pdf

[Zhou15] Zhou, L., and Foster, J. A. "Psychobiotics and the gut–brain axis: In the pursuit of happiness," *Neuropsychiatric Disease and Treatment* (2015): 11, pp. 715–723.

THREATS TO HEALTH

Now that I have described the biological systems that are subject to environmental influence, let's turn to what these environmental influences are. There are a number of factors in life which constitute threats to health and well-being. Some of these are natural, some man-made—we will discuss each of these in turn and then consider how they impact the microbiome.

Agriculture and Food Production

Much of modern agriculture, as we know it, was developed in the aftermath of World War II and the fear of starvation that engendered. Many governments put in place a policy that would make sure that never happened again and food production would be encouraged *above all else*. In other words, quantity would reign over quality and advances of technology would be used to support that agenda. Agriculture would be mechanized, standardized, and technologized—moved away from being a form of husbandry with mixed farms, to one of an industry. I am not a conspiracy theorist—This not so much a conspiracy as a result of many factors converging at one point in time:

- money as the main driver of success (the profit motive)

- the relative powerlessness of individuals compared to corporations

- the imbalance between developing countries and industrialized ones

▪ technology used in the name of progress without sufficient understanding of the complexity of the systems with which we are interfering.

It's a complex picture but the result is the total domination of our food production systems by a few large corporations, whose only motive is profit and a desire to retain that utter dominance *by any means possible* (i.e., an immorality at the heart of the system).

How food is grown and processed affects the microbiome significantly. Routine dosing with antibiotics makes meat a poor choice, unless grass fed and/ or organic. The overuse of them in our food supply (e.g., in animal feeds and so on) has also affected the microbiome of the soil, which then makes it difficult to grow our crops without huge amounts of fertilizer and pesticides. Thus, the only solution really is a return to natural/organic farming which, although popular in some countries, is being systematically destroyed by the agribusiness corporations who want to maintain their monopoly.

In some ways, this mirrors the damage to our microbiome; loss of diversity in our food production via unbalanced mono crop systems has led to lack of resilience to pests without massive dosing with pesticides (equivalent to antibiotics in the microbiome). In the industrialized farming practiced in developed countries, using glyphosate to desiccate crops prior to harvesting decimates our gut lining[1] it destroying the gaps between lining cells (so-called "tight junctions" that control what molecules come in and out of our bloodstream; see Figure 6.1). The semipermeable nature of this junction allows the body to regulate what enters the bloodstream— and this in turn is what balances the immune system. By destroying that essential control system, glyphosate destroys our health over the long term as we become more immune dysregulated, reacting to food proteins as if they were toxins (termed "leaky gut"). Thus, the use of routine spraying with glyphosate is tantamount to chemical poisoning of the food we eat (another reason to go organic in your diet where such spraying is not allowed). Apart from avoiding these toxic chemicals, we need to add specific foods with high inulin (*prebiotic*) content that help to heal the junctions. With such high prebiotic foods like dandelion, artichokes, onion, garlic, and jicama (pronounced hee-kama), leaky gut can be reversed.

[1.] As confirmed by independent scientists as opposed to industry sponsored research

Soil Degradation

There are other problems with this approach. Fifty years of chemical agriculture has denuded the soil to a huge extent. A concentration of high doses of just three minerals (nitrogen, phosphorus, and potassium [also called "NPK fertilizer"] from the chemical symbols for those three minerals) has unbalanced the soil and the routine dosing with herbicides, pesticides, and other chemical additives has killed off the valuable soil microorganisms (bacteria and fungi) that help to maintain a balance in the same way as in our bodies. Hence there are fewer soilborne worms, insects to feed on the worms and then less birds that can live off the insects. It is frightening how many minerals and vitamins our vegetables have lost in sixty years of industrial agriculture [McCance16]. The main loss has been in iron, but magnesium, zinc, and trace minerals have all reduced by 40–60% in sixty years. Our diets have shifted majorly from being mainly vegetables in Victorian times up until WWII, to a majority of meat/cheap carbs now. Not only have the proportions changed, but the vegetables we do eat are lacking the essential nutrients they once obtained from the soil. No wonder we are all deficient *and* yet overfed!

However, when crops are grown in rotated, naturally fertilized soils (the unused portion of the crop is allowed to rot naturally and is ploughed in along with animal manures as in preindustrial agriculture, the soil microbiome balance is restored. I, myself, get an organic vegetable box delivered—it is one of the few luxuries I will allow as I know the vegetables have been dug recently, I don't need to wash or peel them so much to rid myself of the toxic sprayed chemical residues present on regular vegetables.[2] In fact, if I retain some soil on the surface of freshly harvested vegetables, all well and good; the soil itself is an important regulator of our gut flora and needs to be eaten in small amounts as it has been for millennia. Therefore, if you are lucky enough to be able to grow your own without chemicals or can afford to buy organic—don't wash it off so much as it's good for you!

Frankenstein Food: Artificial Foods and Additives

We have entered a new age in which "food products" outnumber real foods in the average supermarket. I make a distinction between these two things; food products are chemically composed/processed foods with a mixture of real ingredients and chemical additives that help extend shelf-life, add

---

[2.] Most fruit and vegetables grown in industrialized countries are sprayed dozens of times with various pesticides and herbicides. This residue often remains in the skins and is not destroyed by cooking. Some are worse than others.

palatability, and sometimes vitamin content. But the processing that the food has undergone to make it supermarket compliant means it bears little resemblance to the food you would make from scratch at home. And the additives have never been tested properly for their toxic effects in the body. They may have been tested on animals to show whether they are overtly carcinogenic but not for the effect of their *cumulative* ingestion. The liver has to rid the body of any chemicals that are not overtly needed for biosynthesis or metabolism; these chemicals are often new to the body in evolutionary terms, and they interact with each other in unpredictable ways.

Ready meals are an innovation that shows the depths to which we have gone in our search for meals that take no time at all to prepare. Many look very appetizing and have gone through extensive testing and recipe tweaking to make them attractive to consumers. However, they are made in factories not kitchens so to keep them fresh on the shelves they will have a battery of preservatives and flavor enhancers added.

My advice would be to avoid these contaminated foods and food products (i.e., anything with more than five to seven ingredients). These days I skip the "interior" aisles of supermarkets and stick to the fruit, vegetables, and some whole foods. But even in the wholefood or gluten-free section it is hard to avoid added sugar, high fructose corn syrup and preservatives. You have to change your shopping habits if you want to eat well. Learning to cook with fresh ingredients and being wise to additives is very important if you want to lower your body's toxic load, especially if you are female (where higher fat levels store more toxins).

GMOs and Hybridization

This is a very controversial area that is full of politics and passion. I think we can all agree that foods have changed significantly in the last forty or fifty years—especially since it became possible to alter the genetics of food crops directly. We have long developed hybrids by crossing one species with another to improve food production. But with scientific advances in DNA modification technologies, we can now introduce genes from one species directly ("splicing") to give the recipient species characteristics from the donor. Hence, we can introduce genes that make a crop resistant to disease, for instance.

Here we are going to look at the special case of wheat and dairy production which illustrates well how technology has intervened, with mixed results, which, according to the influential Weston A. Price Foundation, "provides revealing insights into modern agriculture and food production" [Czapp06, para. 3].

Milk Protein—A1 to A2 Change

During the 1960s, agriculture was undergoing industrialization, in order to feed a growing population more efficiently. Milk production came under particular scrutiny as milk was considered a health food which would "build the nation." Slogans from national TV advertising, which included "full of natural goodness," encouraged people to drink a pint of milk a day for health. Children in the UK had free milk delivered to school[3] to help them "grow healthy bones." There are many problems with this belief, however it probably isn't appropriate to go into here.[4] It is complicated; and like most forms of "mass medication," it fails to take account of individual differences in tolerance. Certainly, I think drinking cow's milk seems to have made successive generations taller and bigger than their forebears. But that is not necessarily a good thing. Musculo-skeletal problems can occur if bodies grow too fast under the influence of artificial growth hormones present in milk.

How did it come to pass that we have altered our milk production so significantly that milk has now become an allergen to many? Why is it that we have industrialized milk production to the detriment of many people's health? The problem with this model was that cows were just not producing enough, so farmers were encouraged to change the type of cow from the traditional Jersey breed to the American Holstein, which has a different type of protein in the milk. Jerseys tend to produce A2 type casein protein while Holsteins contained more A1. This has implications for human health as the two proteins are not treated the same way in the body. "We've got a huge amount of observational evidence that a lot of people can digest the A2 but not the A1," says Keith Woodford, professor of farm management and agribusiness at New Zealand's Lincoln University.

[3.] I remember well throwing the milk down the school sinks as I hated it! I think I must have instinctively known I was lactose intolerant without having a name for it.

[4.] Epidemiology (the study of populations) actually shows a reverse association; the more milk a population drinks, the worse the incidence of osteoporosis. But it is a controversial area that depends on your gut flora balance and the type of milk you drink. In one study of Polish women for instance, more milk = higher bone density. It may be beneficial for some people in low amounts whilst for those of East Asian or African descent, it is likely detrimental.

Ancient breeds like the Jersey that tend to be found in Africa and Asia, are less likely to be intensively managed, and don't produce as much milk—this tends to be tolerated better by human digestive systems. The reason is: when we digest A1 beta-casein, a milk protein produced by most cows in Western herds, we cannot break down the protein chain as well as the older African A2 variant.[5] A1 protein is a mutation that probably occurred about 10,000 years ago for reasons unknown but must have conferred an advantage at the time to suckling calves but to which humans, with our relatively poor digestive power, are not well adapted.

There is much research going on into this now as it seems A2 varieties are linked to reduced occurrence of diabetes and heart disease. Certainly, when I have drank Jersey milk (you can get it in some supermarkets, as well as a few farms that specialize in Jersey milk), I have noticed no lactose intolerance and a different interaction with my gut, that is, it is tolerated without bloating and I have a different smell to my body. Other people have corroborated this effect. This is, of course, only anecdotal evidence and means nothing to science that has embraced the clinical trial as the only source of reliable information. But people often take up ideas by witnessing them in others and trying for themselves. So, although this isn't "evidence-based" information in that sense, I am honoring the much older tradition of observational learning. For more information on the history of milk production, I recommend *The Devil in the Milk* by Keith Woodford [Woodford10], which has been updated with recent research.

Be aware that it is not just the change in cow genetics but also the changes in milk *processing* (pasteurization of milk to make it sterile). These chemical industrial processes are relatively recent in our history (only since postwar milk quotas were introduced) and make the body even more reactive because of its destructive effects on the gut. All these changes to productivity may be combining to make our immune systems more reactive. High productivity comes at a cost to our health.

Increasing Gliadin in Wheat

The same is true in the case of wheat—we have changed the way we make bread very significantly in the last forty to fifty years. Prior to the advent of industrialized production, bread was made by "proving" the dough;

[5.] The difference between A1 and A2 proteins is subtle: They are different forms of beta-casein, a part of the curds (i.e., milk solids) that make up about 30% of the protein content in milk. In fact, the difference is only one amino acid histidine instead of proline at position 67 out of the 209 amino acids in the chain. The effects of this small change are important and may reduce lactose intolerance and diabetes.

this involves kneading the mixture of flour and water with a baker's yeast (consisting of specific microbes that can partially ferment the proteins). This is an age-old process, using natural ferments, that was recycled between batches,[6] it's called *leaven*. It uses natural yeasts to partially digest for us, so our guts then have less work to do. It was abandoned in the early part of the last century as it was too variable for the industrialized production and yeast was replaced by a chemical agent called "baking powder." It does roughly the same job of introducing air into the bread but has none of the natural fermentation value for our guts (although sourdough bread aims to redress that balance a little).

Moreover, in the 1950s geneticists were able to develop hybrid forms of wheat that had a massively increased protein component. Remember, we were fighting a war against *protein malnutrition* (not against obesity as is the case now). By changing the protein expression of the genes, growers could increase profit enormously by producing new wheat varieties with massively increased yields. However, the protein component of wheat that delivers this increase (gluten or, more specifically, the subcomponent gliadin), is changed in the process so that it's hardly recognizable to the body. The molecule is much longer and more difficult to digest.

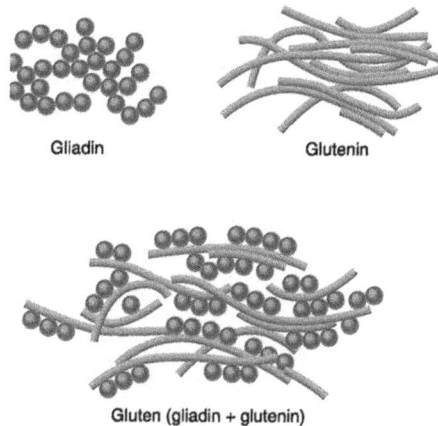

Gliadin   Glutenin

Gluten (gliadin + glutenin)
Figure 3.1. The gluten molecule.

Proteins, like other foods, must be broken into smaller pieces by the action of enzymes in the body. With proteins though, their three-dimensional shape is highly regulated by the particular sequence of amino acids (basic building blocks of protein) in the chain. Like DNA with its code of four

---

[6.] A tradition that is being revived in "sourdough" breads currently available from artisan bakeries and online supplier.

bases, proteins are made of a "code" of a sequence of amino acids; the 3D conformation (shape in space) is determined by this sequence. If you change just a few of the amino acids, the protein shape is altered and that affects the ability of enzymes (which are themselves proteins) to attach and interact with it. Enzymes allow chemical reactions to happen faster and more efficiently via this "lock and key" attachment.[7] If the protein is a different shape, they can't attach properly, and the reaction occurs inefficiently if at all.

The problem seems to be that the molecule, being similar in structure to proteins present on pathogens, is mistaken by the body for an invader and the immune system fights an attack against it in a system of "molecular mimicry." This is because the molecule fits the protein receptor "dock" of the antibody (a part of the learned or adaptive immune system)—well enough to trigger it.[8] However, the immune system in its attempt to rid the body of the "invader," also attacks the lining of the gut (particularly in celiac disease) destroying it in the process. It may also prime the autoimmunity of the thyroid causing Hashimoto's disease [Myers15].

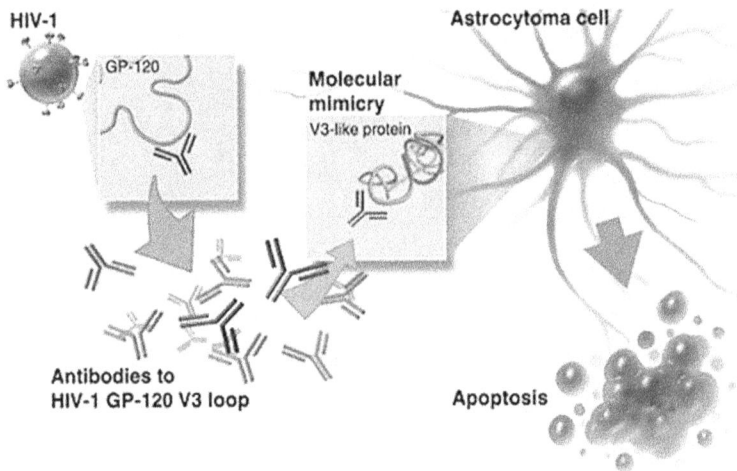

Figure 3.2. Molecular mimicry.

[7.] And possibly also with a quantum electron effect as described earlier.

[8.] The learned immune system is very complex but essentially learns to match the protein "shapes" of invading organisms with the correct protein "dock" or receptor on the antibody which it produces to match it. Once so primed, the immune system T cells are then "primed" to recognize anything that looks similar to it.

Thus, the same thing that happened with milk has happened here; we may have increased yields exponentially but at the cost of our health. Now that we are fighting a war against plenty rather than want, it is time perhaps to return to more natural methods of food processing and perhaps older varieties that were more nutrient dense. This is happening slowly as more and more artisan breads are being offered commercially.

Carbification of the Diet

There has been a massive sea change in the last fifty years in our eating habits. We have gradually come to the point that most of our energy (calories) now comes from carbohydrates as opposed to fat. According to many experts, including science journalist and author Max Lugavare, this "over-carbification" of the diet has disastrous consequences for our ability to handle sugars with the subsequent epidemic rise in diabetes. Diabetes used to be a rare disease. Now it occurs in one in three people; if you include numbers with the prediabetic state of "metabolic syndrome" (aka "syndrome X" in the United States), they are likely even higher.

Moreover, all carbohydrate basically comes down to one word: sugar. The body treats most processed carbohydrate food (e.g., white bread, pasta, sweets, and cakes) as sugar. We know that sugar is bad for us (as is most processed carbohydrate, e.g., white bread has the equivalent glucose "hit" of refined sugar in the body). Sugar when glycated (has certain protein molecules attached) becomes an endotoxin (internal toxin) and promotes aging. In point of fact, we need diets with low amounts of grains, to avoid blood sugar dysfunction. And if we do eat carbs they need to contain "resistant starches" (or slow release such as brown rice rather than white), which take longer to be broken down and so doesn't cause the blood sugar spikes of refined carbs.[9] Some people have had great weight loss by avoiding carbs altogether or grains in particular, but I would never promote taking a whole food group out of your diet. It is all about balance; we need to get more of our energy from fats, as we did in the past.

Most of us are now aware of the problem with sugar and governments in many countries have recently taken steps to introduce a "sugar tax" on the worst culprits, carbonated drinks [Triggle18]. But we seldom hear that sugar also feeds the "bad bugs" in our gut and drives inflammation that further promotes diabetes and Alzheimer's (I talk about that link further in

---

[9.] Interestingly, reheated potatoes become resistant starches compared to freshly cooked ones.

Chapter 5). It is the change in the microbiome that we are going to focus on here. Let's look at the evidence.

Diet and Gut Bacterial Modification

We know from animal (rodent) research that diet significantly changes your gut bacteria. However, human research is less common. What about the reverse idea of changing your gut bacteria to alter your health, that is, as an intervention? Researcher Brenda Watson did an experiment looking at just that; she recorded the bacteria present in different types of people and observed if changing the balance could alter their weight. Her sample of ten people each had a baseline stool analysis, and their weight was recorded, she found that the composition varied significantly depending on your weight. All the larger people had less bacteroidetes, versus firmicutes (a low B/F ratio) despite their exercise regime. So, she slowly changed their microbial balance by exchanging things within their normal diet. They all had to eat three meals a day and two snacks, but they were guided with the following criteria:

- Include small amounts of protein in the snacks—every three hours, one ounce.[10]

- Restrict carbs, especially sugar and some kinds of fruit. Limit to ten teaspoons equivalent per day in an effort to start to balance blood sugar and reduce cravings.

- Improve time management to avoid eating on the run; "be prepared." When you are hungry and you are on the run, you will reach for carbs and sugar.

- Provide a small dish to limit portion size.

- Introduce fermented foods; vegetables (not store bought as they are pasteurized and therefore "dead" bacterially). Also, after three weeks, they added probiotic supplementation of bifidobacterium of a high dose fifty billion cultures.

People in the study didn't count calories or fat, but only teaspoons of sugar to make it easier and less associated with a "diet" in the conventional sense. In this way, they then switched naturally to low-carb foods without feeling a

---

[10.] Snacks were very different to what you or I would consider. It could be a slice of turkey, with a teaspoon of fermented carrot in the middle and rolled up. Or a portion of nonsugared natural yogurt with blueberries, and so on.

sense of restriction. This is the ideal change to strive for. Restriction creates stress in the body; this is the main reason why most calorie control diets fail. When a person goes off the diet, they pile on the pounds as the body thinks it has been in a state of starvation.

Grains were allowed (so not strictly gluten free[11]), but they were encouraged to be of a healthier type, that is, they could have oats, but needed to calculate that within the ten teaspoons a day. The biggest change was a switch to 80% vegetables because fiber is a big component of diet; she found that 25–35 grams of fiber a day is when you get weight loss. Fiber also feeds the friendly bacteria (bacteroidetes), whereas firmicutes feeds on sugar.[12] You might need to add soluble fiber to bring this level up if you can't eat that amount of vegetables (but this is why blending is so useful—more of that later). Stools were then tested every six weeks to check the ratio of bacteroidetes/ firmicute bacteria. They found amazing changes which are summarized in the book "*The Skinny Gut Diet.*" by Brenda Watson. The results are very positive but, in summary, are that you can change your internal microbiome by the foods that you eat, and that, in turn changes your health.

Biofilms

Bacteria exist in several different forms:

- planktonic (single cells moving around independently)—associated with acute infections

- biofilm—more likely in the established colonies of the microbiome

The *biofilm* is defined as a colony of bacteria that are "embedded within a self-produced matrix of extracellular polymeric substance (EPS)" to stick to the surface that they inhabit, as we saw with plaque. They have a purpose in controlling what comes in and out of the body. They can be both good and bad, depending on their constituent bacteria and the context. Biofilms can be considered a colony with a surrounding "fort" of polysaccharides to create a barrier to being attacked. This occurs via a mechanism called "quorum sensing" where the bacteria sense the presence of an antibiotic and form the biofilm as a protective device. This is why "infection often turns out to be untreatable and will develop into a chronic state" [Bjarnsholt13,

---

[11.] Processed gluten-free foods are often just simple carbs anyhow (with often more sugar to make them palatable) so are a junk food with different ingredients!

[12.] One way to remember the ratio, suggests author Brenda Watson, is bacteriodetes = be skinny, and firmicutes = fat bacteria; so the higher your B the better.

Abstract, p.1]; it forms the basis of *antibiotic resistance* (where antibiotics no longer have an effect when used clinically).

Botanicals (certain plant products) interfere with quorum sensing so that bacteria don't know to reproduce or build resistance. Thus, botanicals discourage the formation of biofilms and so the body becomes more efficient at attacking the bacteria [Cosa19].

Conversely, vaccination adjuvants and mercury (from dental amalgam) allow bacteria to form more biofilms so can make the body more likely to develop them. Herbs with hydrolytic enzymes have the ability to break them down and get the mercury/heavy metals out causing a standard *Herxheimer reaction* (feeling ill, having skin eruptions, etc., for example as seen in candida elimination). This can be difficult for some people to tolerate and has to be managed carefully as the solution can often be worse than the problem. Indeed, Dr. Dietrich Klinghardt's research shows that candida is trying to keep the mercury out of the way and could be seen as your friend, therefore. But when it gets out of balance it may cause more problems than it solves. Check your symptoms for possible candida overgrowth but also get tested for heavy metals as clearing one without the other is pointless. There are some great online quizzes to help you ascertain your likely load, and you can order standard tests too.

The ideal way of managing a biofilm is a more gradual "weeding and seeding" approach; knocking out bad bacteria and bring in new "good" types to increase the biodiversity. Once you start to make changes in the gut it alters the biofilms in other parts of the body. This is because the bacteria talk to your brain via their DNA and metabolic wastes that feed other bacteria in different parts of the body. Remember, different bacterial compounds turn our genes on and off via epigenetic mechanisms. Thus, we need to have many different ones to finely tune this. The primary way to control this is via diet; a return to an organic diet with whole and varied foods. The ultimate blueprint is to have a raw, gluten-free diet with fermented foods and prebiotics like root vegetables[13] to promote a longer healthy life. Incidentally, beer used to be a health food, as it was naturally brewed with low alcohol, but sadly that's rarely the case now. Milk can be better tolerated when it is fermented too (as yogurt for instance). We go into this in more detail later.

[13.] According to Tom O'Bryan, author of The Autoimmune Fix, a GFD is missing the important prebiotic qualities of wheat (78% of the food for good bacteria called arabinose islands) so you need to have alternatives like root vegetables, bananas, and asparagus.

Over-Cleanliness

Another big issue in most households at the moment is over-cleanliness. Since the 1950s there has been a gradual push toward chemical cleaners/ detergents/body care products in the home. This is not in our best interests as far as health is concerned. If you look back at a Victorian classic, *Mrs. Beeton's Book of Household Management*, you will note there are perfectly good natural alternatives for cleaning such as lemon juice, vinegar, and various natural soaps. Most other products are made by chemical companies and are full of untested ingredients. People assume that because something is on sale to the general public, it must be safe, but this is not true. In fact, most of the ingredients in cleaners have *never been tested independently* or in combination with each other as they are in most homes. This is shocking but true. The toxins they contain may be carcinogenic in high doses and, although there are generally low levels in them, no one looks at the *cumulative burden* of toxicity over a lifetime of use.

Some people are also just more sensitive to chemicals and this is because their natural detoxification processes do not work as well; this is genetically controlled to some extent by the function of your liver enzymes. We cover this in more depth in the section on genetics. If you are unlucky enough to inherit the poor converter enzyme combination, then it is likely that strong smells, paint fumes, and some perfumes will give you an instant headache, where levels inhaled build up in your bloodstream as your liver is unable to detoxify in time. I had this problem as does my mother.[14] I cannot stand to be in a room or a car with someone wearing strong perfume, and I have to ask them to open the window or move away.

We have long known that we are too clean, and that over-cleanliness is causing an increase of allergies in children [DailyTelegraph10]. It is best to avoid antibacterial wipes, chopping boards made of wood,[15] and cleaning materials as they wipe out all bacteria, not just the bad ones. They are more damaging than beneficial. Use natural products when possible.

Environmental Toxins

We are subject to an unprecedented number of toxins in the environment, whether in the air we breathe, the clothes we wear, the water we drink, or

---

[14.] It is also tied in with poor glutathione detox systems (controlled via a genetic SNP).

[15.] Wood has natural antibacterial compounds in it so is a great material for chopping boards.

the food we eat. Never before in the history of life on this planet have there been so many new chemicals introduced, previously unknown to life. This causes a potential problem as there are few microbes that have evolved that can digest or deactivate these chemicals. Of course, that is changing through mutation and natural selection, but it is a slower process than the rate at which we are releasing new products into the environment.

Hazardous Chemicals in the Environment

Plasticizers like bisphenol A (BPA), a common component of flexible plastic (as in plastic bottles), disrupts our body chemistry directly to form cancer cells (*carcinogen*). It also distorts the body's hormonal function and is therefore a *xenoestrogen*. You may have already been exhorted to avoid storing water or oil in plastic bottles; this is for a good reason. Oils stored in plastic interact with the plastic—the chemistry of oils and plastics are remarkably similar.[16] Decant all oils into glass bottles when you buy them and do not store old oil—only buy what you can reasonably use before the "use by" date—this is especially important in the case of polyunsaturated oils (e.g., olive oil) because they are less stable in heat.

Other chemicals to watch out for are dioxins, phthalates (in personal care products), fire retardants, organophosphates, and glyphosate. You cannot live in a twenty-first-century world and be immune to these problems. Even the Inuit, a group of relatively isolated tribes, have one of the highest heavy metal toxicities from eating shark meat. What goes in the water goes into us. Moreover, fat tissue is a dumping ground for many of these chemicals; hence breast tissue can be most affected. It's one of the hidden causes of the rise in breast cancer that no one talks about.

One thing you can definitely start with is to avoid excess food packaging—getting loose food if possible, and if you have to buy things in packaging be sure to check if it is BPA free (and recyclable). Be particularly careful around soft plastics and fatty foods; the chemicals in the plastic will leach into the fat—try to get them in paper or glass. And absolutely *stop heating food in plastic containers* (even microwaveable—it might say on the packaging that it is "heat proof" or safe, but in fact there has been very little research on the long-term effects of heating such plastics). It is safer to decant the food into glass, stainless steel, or enamel containers (depending on whether you're using conventional or microwave ovens, of course). Avoid nonstick pans, particularly if they are old and damaged. The

---

16. See http://www.plasticseurope.org/what-is-plastic/how-plastic-is-made.aspx.

heavy metal content seems to accumulate in the food and therefore you will ingest it gradually.

Most people (even newborn babies) have a considerable amount of chemical toxicity. With babies, it is passed on through the mother's milk when they breastfeed. Mothers offload their toxicity into the baby as a natural result of concentration up the food chain—that is why the first-born children will often get the lion's share. A typical list of contaminants would be: DDT, PCBs, trichloroethylene, perchlorate, dibenzofurans, mercury, lead, benzene, and arsenic. According to an illuminating article in *The Guardian*, "Because breasts store fat, they store toxic, fat-loving chemicals" [Williams12]. How come these chemicals are present? The answer is surprising. One of the worst places for chemicals is the home. Think about it; it's the place you often spend the most time and yet it is full of the chemical products that advertising persuades us we need: plug in room deodorizers, perfumes, flame-retardants (on our sofas and carpets), dry-cleaning fluids, toilet deodorizers, cosmetic additives, cleaning products, gasoline by-products (Vaseline for instance), home pesticides, and decorating products like paint thinners and wood preservatives. I could go on. We have largely replaced natural cleaning products for nice-smelling chemical substitutes that make us feel good about ourselves in having a "clean" (read sterile) home. This involves using a mass of products for different purposes, some of which are full of largely untested chemicals. It is still a shock to most people that the vast majority are not tested properly for environmental effects—and certainly no one has considered how they accumulate and interact in us. According to latest research as reported in the same *Guardian* article:

> Your breast milk tells the decades-old story of your diet, your neighborhood, and, increasingly, your household decor. Remember that old college futon? It's there. That paint in your bathroom? There. The chemical cloud your landlord used to kill cockroaches? Yup. Ditto, the mercury in last week's sushi, the benzene from your petrol station, the perfluorooctanoic acid (an antigrease coating) from your latte cup and sofa upholstery, the preservative parabens from your face-cream, the chromium from your nearby smokestack. [Williams12, para. 15]

If the mother is deficient in vitamins, antioxidants and good bacteria, the ability of breast milk to combat these toxins is diminished. In a healthy woman breast milk has a huge number of positive constituents to counteract the negative; mostly microbes—up to 600 different species of them.

The report also states:

> Scientists used to think breast milk was sterile, like urine. But it's more like cultured yogurt, with lots of live bacteria. One leading theory is they act as a vaccine, inoculating the infant gut. A milk sample has anywhere from 1 to 600 species of bacteria. Most are new to science. [Williams12, para. 7]

In addition, some of the sugars it contains cannot be digested by the infant and are there purely to feed beneficial bacteria (prebiotics). So, there must be a reason that the bacteria are being encouraged. We now know that these bacteria help us to dispose of toxins, either directly or by creating minerals and vitamins that help to bind them and excrete them from the body. A failure of this process—breast feeding or the ability of the child to digest will impact their toxic load potentially causing mental health conditions like ADHD, autism, and anxiety later on (the brain is also particularly fatty and therefore stores toxins). It's a ticking time bomb that our biology has failed to keep pace with our environmental degradation.

Let's look at each toxin group in turn now.

Heavy Metals: Mercury and Lead

Mercury is increasingly found in fish, like tuna, that are higher up the food chain. It is the second most poisonous chemical known to man—especially when it combines with organic chemicals in the body to form methyl mercury compounds. Tuna is a popular fish in human diets and is also eaten by larger fish, so the higher up the food chain the fish we eat, the more likely it will be to carry a heavy toxic load. The other main method we ingest mercury, is via "silver" dental fillings which I've already covered in the section on the dental microbiome. It is unbelievable that we would put this compound in our mouths with a wet tissue (our gums) surrounding it to act as a battery. But, given that it is still accepted practice within many countries including the United States and UK, we have to find ways to either avoid new amalgam fillings or mitigate the effects somehow if we have already received some. To do this we use oral *chelation* (pronounced key-lation) with minerals such as Selenium (Se), Zinc (Zn), cruciferous vegetables, milk thistle (a herbal tincture), dimercaptosuccinic acid (DMSA), and chlorella (a type of algae). This is discussed in more detail in Chapter 4.

Lead and mercury at chronic low levels are often found in old houses too—my exposure has been both through dental work and paint. As a child, I used to love to paint pictures with a little lead-tipped brush which I would often lick before applying the paint from the lead tube. Then, as a teenager, I used to do odd-jobbing and once stripped an entire house of its old (lead-based) gloss paint with a blowtorch (!!). Plus, probably the most worryingly, I worked in my father's factory in the summer holidays; it was a metallurgy workshop. I worked sorting lead and stainless-steel filings, and the air was thick with metallic dust. My father would bring it home with him every day too, of course. A triple whammy for my young body.

There is a theory, promoted by health writers such as Patrick Holford in the UK, that issues such as Reynaud's syndrome (white finger) is the body's way of diverting blood away from the most toxic tissues this would make sense in my case as it is my right index finger that is affected and no others. But whatever the reason, it is a common problem linked to stress, with toxicity as part of the picture; but there are links with autoimmune diseases generally, so the picture is complex.[17] I would encourage you to get tested if you are at all concerned and follow a detoxification protocol if levels are high.

Pharmaceuticals

Drugs, especially antibiotics, food, and toxins all alter the gut and body flora with cumulative effects; what author Dr. Mark Hyman MD has called a "perfect storm" of changes. Let's focus on the first of these.

Medications (pharmaceuticals mostly) are an important epigenetic modifier of the microbiome, albeit in a negative, unregulated way. Here are the main culprits:

- antibiotics (including those used in the food supply)

- acid-blockers/proton pump inhibitors (these change the acidity of the stomach and therefore alter the pH in the rest of the digestive tract)

- weight-loss drugs and cholesterol-lowering statins

- birth control pills (these upset underlying hormonal issues)

- diuretics/antihypertensives

- NSAIDS (non steroidal anti inflammatories - studies have found they increase the risk of heart attack) [Arfe16]

Antibiotics are like a nuclear bomb and cause profound changes in the microbiome, aside from their intended target which would be a pathogenic

---

[17.] See http://www.innovativetherapycanada.com/show.raynauds.html for more info.

bacterium somewhere else in the body, for example in the case of meningitis. From their routine use in everything from minor infections, prophylactic use in dentistry, veterinary medicine, and animal feeds, they are slowly but surely reducing the diversity of the gut flora and thus our resilience to chronic disease.

While they are undoubtedly one of the miracles of twentieth century medicine that have enabled us to survive many infectious diseases that used to routinely kill us, the legacy is one that may yet wipe us out, albeit much more slowly. With the advent of microbial resistance, the microbes are biting back. With their immense capability to mutate and swap genes around, these microbial survivors from antibiotic treatment become ever more resistant and become the "superbugs" for which we have no current answer. Without wishing to sound doom-laden this could be one of the biggest challenges yet to our survival. Even governments such as the UK are recognizing this [DOH13].

Don't forget, there are not only the elective ones you take for sickness but the ones you have no control over that are in the food chain. The use of veterinary medicine is endemic; the reliance on them to sustain unhealthy animals, kept in unnatural conditions has many similarities to the growth of monoculture foods in our agricultural system. In both cases we need a reliance on artificial chemicals to maintain a system that is basically unhealthy. This is madness. Antibiotics have direct effects on our health too. The antibiotic, tetracycline, for example, can block absorption by binding with minerals, such as calcium, magnesium, iron, and zinc in the GI tract.

Weight-loss drugs and cholesterol-lowering medicines similarly bind to fats, preventing them from being absorbed. Drugs that treat acid reflux or heartburn raise the pH environment of the upper GI tract, which reduces absorption of needed vitamins and minerals. This is especially problematic among the elderly, who often are already low in stomach acid.

Nutrients are essential to the metabolic activities of every cell in the body. They are used up in the process and need to be replaced by new nutrients in food or supplements. Some drugs deplete nutrients by speeding up this metabolic rate. These drugs include antibiotics (including penicillin and gentamicin), steroids such as prednisone, and gout medication such as colchicine.

Other drugs block the nutrient's effects or production at the cellular level. In addition to the intended effect on enzymes or receptors, medications can influence enzymes or receptors that help process essential nutrients. For example, widely prescribed statin drugs block the activity of hydroxymethylglutaryl coenzyme Q10 (HMG-CoA), an enzyme that is required to manufacture cholesterol in the body. This action also depletes the body of the important protein molecule coenzyme Q10 (CoQ10), which requires HMG-CoA for its production. This has a serious negative impact on muscle and heart health. Anyone taking statins should also take CoQ10 as a supplement. In some countries, this is now routinely recommended although not in the UK [UKMi16].

Drugs also can increase the loss of nutrients through the urinary system. Any diuretic drug can drain the body's levels of water-soluble nutrients, including B vitamins and minerals, such as magnesium and potassium. The major offenders are medications to treat hypertension, particularly diuretics that reduce blood pressure by increasing the volume of water flushed out of the body.

Medicalized Childbirth

Although I go into much more detail on this in the next chapter where we look at the life cycle, I want to summarize here that the rise in surgical births and lack of breast feeding are having rising catastrophic effects on the microbiome.

As we pass along our bacteria as well as our genes to our children, our imbalance is passed on too. As mothers have found, the birth process has been gradually wrested from them with the advent of surgical birth practices like amniocentesis, epidurals, assisted birth (inducing forceps, suction, etc.) and, of course, more surgical interventions like C-sections. What should be a private, personal, dare I say, *spiritual* process between mother and child (and her attendant family and midwife), has become a full-scale medical intervention with all the fear and lack of control that it implies. It can be a huge trauma to both mother and baby to be surrounded by masked gowned people with machines and bright lights. There has been much research recently on how much a newborn baby (or even fetus) can indeed feel pain and distress along with the mother.[18] Most women seem to manage well enough, if they are prepared well and supported, but for some, especially if there are complications, it can be highly traumatic and the effects on the child are also profound.

[18.] The burgeoning field of pre- and perinatal psychology (PPN) is overturning of lot of previously held views on babies and their ability to feel.

Transpersonal psychological work developed by Stan Grof [Grof85], has shown that adults retain a memory of their birth even if not consciously remembered, and it can have huge effects on their adult functioning. For instance, if they were delayed coming through the birth canal, there is a lot more fear and stress hormones circulating for longer and those children are often very stressed when they eventually come out into the world. If these effects are not immediately mitigated by being cuddled and held by the mother (with the smell, warmth, sound of her heartbeat and voice reminding the newborn of its previous place of relative comfort), then the baby may well have a memory of danger (threat) that persists into adulthood. This memory can be unconsciously recalled whenever a stressful event occurs in adulthood—we call them sensitized—and this abnormal stress response diverts the autonomic nervous system into disabling normal gut function. Hence, children will often display digestive disorders early on, which can develop into more serious illness later if their subsequent childhood is also stressful.[19]

There is much to be said for redesigning the birth process away from this clinical idea and incorporating elements of nurture and connection; the use of doulas and "hypnobirthing" (the use of hypnotherapy to relax the stress response) point the way to a more holistic experience which many mothers yearn for. Of course, there will always be situations where surgical intervention like emergency Caesarian is required and sometimes this will save the baby's or mother's life, but for most it is excessive. The effects of such surgical births (where the baby is pulled out of the womb rather than inoculated by passage through the birth canal) on the microbiome of the baby are too important to ignore.

Vaccines

Vaccinations have been in use for nearly 200 years; their basic principle is to prime the acquired immune system to recognize a pathogen by the injection of semilive/dead bacteria directly into the blood stream. The immune system then produces antibodies to the pathogen that prime it for any future interaction with live bacteria. There has been great controversy surrounding the use of vaccines. For some, they are the white heat of progress; they have indeed allowed us to conquer most of the infectious diseases that children and adults routinely died of, such as cholera, diphtheria, tuberculosis, and so on.

[19.] This was the premise of my book *The Scar That Won't Heal*.

The problem seems to be with the small number of children who have reacted badly to the high doses of live bacteria/adjuvants in some of the more complex combination ("multivalent") vaccines like the MMR (measles, mumps, and rubella) vaccine. Standard adjuvants used include mercury (as thimerosal) and, more recently, aluminum, but the list of other commonly used ingredients is sobering: propylene glycol (antifreeze), formaldehyde, glyphosate (weedkiller), MSG, and certain emulsifiers like polysorbate 80 that breaks down the gut lining and blood brain barrier (necessary for all of us to protect the brain) allowing toxic chemicals to get into the brain directly. The idea of such adjuvants is to hyperstimulate the immune system to provoke a more significant response. No account is taken of size/weight/sex or microbiome; all babies are given the same dose. The other problem with this approach is that the number of vaccines given to very young babies has increased markedly in thirty years;[20] up to triple the dose in the United States, for instance, although in Europe we have fared marginally better. In Japan, the MMR vaccine has been discontinued and the single mumps and measles shots are given instead, as the evidence shows the young immune system better supports them when they are *not used in combination*.

Without getting into the muddy waters of this debate (this has been done better elsewhere[21]), I do believe there is a problem with *some* children whose immature immune systems are not able to cope with the onslaught. It can cause a variety of damage; with autism being the most controversial. The issue is difficult to debate without getting attacked for lack of scientific proof. Because it only happens in a few cases, it is difficult to prove as the primary causative factor. Moreover, of course, knowing what we now know about the microbiome, it clearly isn't the only factor. The resilience of the baby's gut flora would be another factor which would be a variable that cannot be predicted ahead of time.

But the issue doesn't seem to go away and will be debated further, I am sure. For an informed choice you need to read more widely than the

---

[20.] According to most sources, approximately three time the amount is given to US children now as compared to in 1950. http://vactruth.com/vaccine-schedule/. In the UK it is perhaps less of an increase, but still significant; see http://www.chop.edu/centers-programs/vaccine-education-center/vaccine-history/developments-by-year.

[21.] For a very research-based impartial viewpoint, see Dr. John Campbell on his YouTube channel https://www.youtube.com/@Campbellteaching, or for more controversy see Del Bigtree on https://thehighwire.com/ and also London Real https://londonreal.tv/ who cover a lot of vaccine-related topics since the pandemic.

standard medical literature. A book I highly recommend is *"Vaccines: A Parent's Guide"* by Richard Halvorsen [Halvorsen13], which manages to synthesize the arguments well without patronizing those parents who have concerns about vaccinations or by totally denigrating their use as some groups have done. There continues to be debate about their safety and efficacy [Mold16]; it is a very political arena with many vested interests at stake, especially since the COVID-19 pandemic. Tread wisely and carefully and whatever you do ask your doctor to read you the list of ingredients in any vaccine and known side effects before you agree to vaccinate your child. This is the true basis of informed consent. Without it, you are flying blind basing your judgment on what you have been told and this is not always accurate. Your GP may not even know enough to give you full informed consent, and thus is complicit but that's not to blame them as individuals. They have also been persuaded by a very active and moneyed lobby to ignore any doubts and trust the pharmaceutical industry that sponsors their training.

Parasites

Most of you will be unaware of parasites as a factor in health; so, it seems, are most medical doctors. It has been claimed, for instance, that 8/10 Americans are infected with one or more parasites; no doubt a similar proportion in the UK. It seems that it is just too uncomfortable a subject to discuss. Yet within veterinary medicine it is well known and accepted; we de-worm our pets regularly. How can it be that our domestic pets that live with us have parasites, but we do not? It's impossible—we transfer eggs between them and us, and we come into contact with other eggs via our interaction with the soil and food. Today we do have more ability to detect them than ever before so hopefully this ignorance will change.

One of the most common infestations is yeast overgrowth (Candida species). Not strictly a parasite as it is a natural part of the human gut, it is a fungus that can overgrow. According to expert Dr. Heinrich Kramer, antibiotics cause "increased colonization of fungi (Candida albicans) in the gut which, in the process of mutation, form roots, change their metabolism and secrete toxins" [Conti09]. Overgrowth will become a problem as the threadlike *mycorrhizal form* is able to penetrate the gut wall and make it leaky, thus increasing the likelihood of autoimmunity. The aim, therefore, must be not to eradicate it entirely but to *treat the imbalance*. If you do this, indigestion will often disappear. Candida constitutes one of the most immunosuppressive agents in the human body, although there are still more being detected.

Symptoms

The many and varied symptoms of intestinal parasites are bewildering; a UK expert, Emma Lane, has talked of a growing awareness that parasites may be behind many chronic conditions [Lane17] such as:

- diarrhea, bloating or gas—not otherwise diagnosed, that is, not due to food sensitivity or autoimmune problems

- GERD—(gastro-esophageal reflux disease) caused by abdominal pressure on stomach.

- Hiatal hernia (caused by weakened esophageal muscle)

- Crohn's disease

Other symptoms of parasites may be as diverse as:

- difficulty sleeping—waking up between 1 and 3am

- frequent infections, cold or yeast infections of the skin

- sudden weight gain or loss.

- skin conditions like rashes, hives, eczema, psoriasis, or Seborrhea

- respiratory illness, for example, asthma (can be caused by roundworms passing through the lungs)

- chronic fatigue and fibromyalgia sufferers may have associated infections of Giardia

- autoimmune disease (AID), anxiety, depression, even cancer

- joint or muscle pain can be caused by amoeba in joints (which has been hypothesized was the sole "cause" of Rheumatoid Arthritis (RA) [Diana12]—other researchers have implicated candida and mycoplasma

However, it is worth noting that many diseases are *accompanied by parasites*; this is not necessarily causative. But certainly, paying attention to internal cleansing (e.g., with herbs) will help maintain your health even if a specific parasite hasn't been identified. All parasites exist to feed off you; they take your nutrients, specifically blocking absorption of Vitamin B12, fatty acids, and certain important minerals. You would do well to have an annual liver/colon cleanse in any case.

Testing

One can check for parasites and types of bacteria via stool testing. We don't routinely check for parasites in western nations and, as there may be different expressions in Western nations, routine stool tests don't always pick them up as most labs are not particularly specific and therefore accurate. This is not the fault of the labs as such. The problem is parasites, like Helicobacter pylori (H. pylori) and Clostridium difficile (C. diff), live in the intestinal wall or small intestine and so are difficult to identify accurately as they evade the usual markers which don't penetrate the gut wall but stay in the bloodstream.

Giardia and amoeba (commonly acquired by eating out in unclean restaurants and from street food) can be more easily detected but there are no perfect tests, so one also needs to look at symptoms like anal itching, unexplained fatigue, and so on. If standard treatments don't resolve the symptoms, then suspect parasites—and if they are identified you will need to test family members as well. Clostridium, however, like the yeast candida, is a normal commensal that, in balance, helps the mucus lining of the gut stay intact, so shouldn't be thought of as a "bad" bacterium per se. The problem is in overgrowth/imbalance, as always.

If you have traveled recently and have developed sudden symptoms, suspect parasites. There are now online labs that will identify them for you, or you can consult a nutritional therapist who will be able to organize testing for you. Unfortunately, your doctor will likely dismiss such an idea unless you've been in an area known to harbor them, for example, sub-Saharan Africa. Routine parasite infections are not a commonly taught subject in medical school and so your doctor is unlikely to know. We will look in the next chapter how to eliminate parasites if they have been identified.

Stress and Emotional Health

In my previous book, I looked in detail at the stress response and how it relates to physical and mental health. I don't intend to go over it in detail here. But there are some important principles to understand as they relate to the microbiome:

- Stress is anything that overwhelms our capacity to cope, that is, takes the body out of homeostasis, for example, toxins, trauma, and negative beliefs that perpetuate the hormonal cascade. It can become chronic when it becomes a fixed response to threat. This is largely a learned response.

▪ The stress response can be programmed through early life (and some would argue prebirth) experience so that it is easily kindled in the brain and body. This instills a habit of being, such that anxiety and depression become much more likely.

▪ Early emotional trauma such as attachment issues (e.g., an unloving, depressed, or emotionally disconnected mother), accidents, bullying, or ill health landscapes the brain for more trauma later in life. This traumatic memory is stored in the unconscious parts of the brain as a survival response that is easily retriggered and can cause myriad seemingly unexplained medical symptoms throughout the body. The most common, though, are chronic pain, digestive issues, and skin complaints.

It is impossible to separate the mind and body, as the two are in constant communication via the microbiome, so to say something is a physical problem without considering the emotional/psychological issues is not a valid response. We are whole beings with complex mind and body interactions. We must treat all illness holistically, i.e., with both components related to the person's individual life experience.

Stress Response in Health and Disease

Stress is often missed out altogether in the discussion of health and disease, but it is a major modifier of the brain, changing both its plasticity (adaptability) and development. *Neuroplasticity* is the term given to the modification of neuronal pathways (the chain of nerve cells) as a result of changes in the environment. It allows the brain to adapt and grow its neuronal connections, dependent on the environmental needs of the growing child. This plasticity happens most acutely in the first years of life when a baby/toddler is rapidly adapting to its surroundings/learning environment. If the parents are supportive, the brain develops normally with good connections, stress is mitigated (although it can't be avoided, the child learns to self-soothe so grows up with a normal tolerance), and a sense of self is engendered which stands the child in good stead for their future health and well-being. But this can go wrong at any stage if the parental/family situation is a stressful one with lifelong effects on *resilience*—the ability of that child to withstand stress. They develop a low stress tolerance and hyper-reactivity. This is covered in more detail in my previous book *The Scar that Won't Heal.*

Moreover, neuroplasticity is not limited to childhood; in fact, recent imaging studies of the brain have given us a very clear idea that the brain, particularly areas like the hippocampus, are able to change dynamically throughout its lifetime.

> With the spectacular advance in brain imaging technology (particularly via functional MRI (fMRI)), we can see that neuronal pathways are constantly changing—new neurons being formed in areas of high usage and new neural connections lighting up when we start to think or behave in novel ways. So, the brain is not a static transmitter but an organic filtering station, constantly responding to input and altering its function to suit. There is a lovely phrase used in neuroscience— "what fires together wires together," which means that when certain pathways light up the chain of neurons becomes more established and easier to fire next time. In other words, the brain is adapting to its environment and what is needed to make the process more energy efficient and more automatic." [Worby15, p. 113]

The impact of childhood stress upon the developing microbiome is less well-known; we know that children generally are not getting the microbial diversity from their mothers or soil-borne flora, but the cause remains unclear. Stress-related disease in children (allergies, migraines, and chronic fatigue) is rising inexorably year upon year. Stress hormones produced in the early years change gut flora significantly toward more imbalanced neurotransmitter production (remember most of these are made in the gut). Stress impacts the gut hugely to directly turn certain genes on and off (*epigenetics*), but it is the bacterial changes which have the most effect on long-term health. A recent study in mice showed that "the interaction of bacteria and early-life stress may be what determines an individual's likelihood of developing anxiety and depression" [Gregoire15, para. 10]. The authors studied normal mice and "germ-free" mice without gut bacteria. Both produced high levels of the stress hormone cortisol, but only those mice with normal gut flora (not the germ free) showed signs of depression and anxiety which suggests that changes in gut flora and resultant changes in neurotransmitter balance affects mood.

Managing Chronic Stress

Stress is an enormous factor in health and well-being. Stress is much misunderstood; it is not simply a temporary imbalance—if chronic and endemic, it can annihilate your microbiome. Remember, your decisions of the past are the architects of the present [Bock17] if you do nothing to address them. We need to recognize and then manage the broad array of stresses that we are under. It is not as simple as avoiding stressful situations.

First, many of them will be *unconscious*, and second some of them will be unavoidable.

An understanding of stress becomes helpful as you learn to choose your response to (particularly habitual/unconscious) stress, so that you can fight your battles wisely. The stress response, as we have seen, is largely a *learned response* developed from some innate systems and accentuated by particular experience and personality types. The balance of the microbiome plays into this as it helps us signal our internal milieu; their health is linked with ours. Therefore, we need to support ourselves (and our microbiome) by overwriting the stress message with new habits of relaxation such as massage, meditation, and saunas. Learning to appreciate the importance of your gut bacteria and our interconnectedness sounds simple enough. But how do you actually achieve this? Chapter 4 addresses this in more detail.

References

[Arfe16] Arfe, A., et al. (2016). Non-steroidal anti-inflammatory drugs and risk of heart failure in four European countries: Nested case-control study, *BMJ* (2016): *354*: i4857.

[Bjarnsholt13] Bjarnsholt T. The role of bacterial biofilms in chronic infections, APMIS Supplement (2013): 136, pp. 1–51.

[Conti09] Conti H. R., Shen, F., Nayyar, N., Stocum E., Sun, J. N., Lindemann, M. J., Ho, A. W., Hoda Hai, J., Yu, J. Y., Jung J-W, Scott, G. F., Masso-Welch, P., Edgerton, M., and Gaffen, S. L. "Th17 cells and IL-17 receptor signaling are essential for mucosal host defense against oral candidiasis," *Journal of Experimental Medicine.* (2009): *206*(2): pp. 299–311.

[Cosa 19] Cosa, S., Chaudhary, S. K., Chen, W., Combrinck, S., and Viljoen, A. "Exploring common culinary herbs and spices as potential anti-quorum sensing agents," *Nutrients* (2019): *11*(4), p. 739.

[Czapp06] "Against the grain." Weston A. Price Foundation, July 16, 2006. http://www.westonaprice.org/modern-diseases/against-the-grain/

[DailyTelegraph10] "Excessive cleanliness to blame for allergy rise." *The Daily Telegraph*, April 15, 2010. https://www.telegraph.co.uk/news/health/news/7589193/Excessive-cleanliness-to-blame-for-allergy-rise.html

[Diana12] Diana. "The many causes of rheumatoid arthritis explored." My RA Diary, December 8, 2012. http://www.myradiary.com/170/the-many-causes-of-rheumatoid-arthritis-explored

[DOH13] Department of Health and Social Care. "UK 5 year antimicrobial resistance strategy 2013 to 2018." GOV.UK, September 10, 2013. https://www.gov.uk/government/publications/uk-5-year-antimicrobial-resistance-strategy-2013-to-2018

[Gregoire15] Gregoire, C. "How early-life stress could increase risk of anxiety and depression later in life." *Huffington Post*, July 30, 2015. http://www.huffingtonpost.com/entry/gut-bacteria-mental-health-connection_us_55b8d6d6e4b0a13f9d1ade20

[Grof85] Grof, S. Beyond the Brain; Birth, Death, and Transcendence in Psychotherapy. New York Press, 1985.

[Halvorsen13] Halvorsen, R. *Vaccines: A Parent's Guide* (2nd edition). Gibson Square Books Ltd, 2013.

[Lane17] Lane, E. "The Parasites Within: How to Deal with Unwanted Guests.", IHCAN, 2018. https://www.ihcanconferences.co.uk/parasiteswithinpcieurope/.

[McCance16] McCance, R. A., and Widdowson, E. M. (2016). *Cytoplan manual*. Cytoplan UK, 2016.

[Mold16] Mold M., Shardlow E., and Exley, C. Insight into the cellular fate and toxicity of aluminium adjuvants used in clinically approved human vaccinations. *Scientific Reports*, (2016): 6(1), 31578.

[Myers15] Myers, A. "The Gluten, Gut, and Thyroid Connection." Amy Myers MD, January 27, 2015. http://www.amymyersmd.com/2015/07/the-gluten-gut-and-thyroid-connection/.

[Triggle18] Triggle, N. "Soft drink sugar tax starts, but will it work? BBC News, April 6, 2018. https://www.bbc.co.uk/news/health-43659124

[UKMi16] "Should patients on statins take Coenzyme Q10 supplementation to reduce the risk of statin-induced myopathy?" UK Medicines Information, November 2016. https://sandwellandwestbhamccgformulary.nhs.uk/docs/Co_enzyme_Q_and_statins-FINAL%20(3).pdf

[Williams12] Williams, F. "The wonder of breasts." *The Guardian*, June 16, 2012. https://www.theguardian.com/lifeandstyle/2012/jun/16/breasts-breastfeeding-milk-florence-williams

[Woodford10] Woodford, K. *The Devil in the Milk*, Craig Potton Publishing, 2010.

[Worby15] Worby, P. *The Scar that Won't Heal.* CreateSpace, 2015.

CHAPTER 4

FOUNDATIONS OF HEALTH: NUTRITION

So now that we have established the importance of the microbiome throughout our bodies, looked at the threats to their integrity and how metabolic function is tuned to the environment, we need to turn our attention to gut health solutions and the foundations of health.

Good Food as the Foundation of Health

Eating good healthy food is all about creating the right environment for your gut flora so it can make you healthy. With the inexorable rise in obesity, it is good to know that your gut flora can help to regulate your weight because it regulates metabolism and your appetite. Therefore, changing your diet will change the flora, and that will change your metabolism. We will look later at how to "fertilize and create a healthy inner garden," to use a gardening metaphor, but as a quick guide we do this by reducing/cutting out refined carbohydrates and sugars, excess meat, and damaged processed fats; all of which imbalance our flora.

Fresh food, provided it is grown in good soil, with access to light and clean air and water, not only gives us the nutrients we need to survive but it also delivers the beneficial bacteria in the correct formulation via its link with the microbiome of the soil. These microbes then help us create the nutrients we can't make ourselves (or in sufficient number) to be healthy.

The type of food is key: anything that creates systemic inflammation will undoubtedly create poor health outcomes. This includes sugar, dairy, and gluten for most people. Sugar is particularly bad, not just because of its addictive qualities and high calorific value; it also combines with the proteins (and some lipids) on the surface of every cell to produce *advanced glycation end* (AGE) products, appropriately shortened to "AGE's." These "play an important role in the pathogenesis of diabetic complications" [Singh14] in particular and are the cause of skin aging [Gkogkolou12], especially liver spots on the hands and face (with sunshine exposure the AGE's go dark brown/black), but, more worryingly, they are linked to aging of blood vessels with heart disease implications.

Fat is not the Enemy

We have grown up with the idea that low fat diets are good for you. This is very unfortunate as it is, in fact, not true. It is an oversimplification to say all fats are bad. Not all fats are the same. We need to get our fuel from natural fats not processed (hydrogenated) oils. We need certain fats, for example, omega 3 fatty acids that we used to get from eating fish and grass-fed animals (cows are mostly fed artificial fertilized grain-based feeds now). We used to especially prize grass-fed cows' produce (cream and butter, particularly from spring grass which is high in omega-3), but with the anti-fat drive and the rise in lactose intolerance it has been replaced largely with processed margarine and low-fat spreads (none of which are natural fats despite what they say on the packaging). Beware any food product which has a health claim unless it has just one or two ingredients.[1] It is unlikely to be a real food.

There is a very well-developed science around fats, however I don't have time to go into here. I refer you to some of the many good books on the subject like *Know Your Fats* by Mary Enig [Enig00] if you are interested. However, one thing you should be aware of is the importance of short-chain fatty acids (SCFAs) and medium-chain fatty acids (MCFAs) like butyrate contained in ghee (clarified butter) and butter. These are powerfully antibacterial (i.e., they fight against the predominance of pathogenic bacteria), while helping to maintain the gut lining as they encourage the healthy bacterial microbiome. According to a recent study by scientists in Mexico, butyrate has many health-giving benefits mediated through the microbiome. This includes "the ability to enhance the growth of Lactobacillus and Bifidobacterium in the colon and it also has various

---

[1.] Nuts and seeds would be a good example of a good food– the natural fats they contain are good for you. The type of food I am referring to here is a margarine that has been modified to have specific qualities. Unfortunately, they seldom improve upon nature and are often highly processed.

beneficial metabolic effects such as improving thermogenesis and energy expenditure, which contributes to reducing body weight and other markers of metabolic syndrome" [Juárez15, para. 7].

Butyrate is a short six-carbon chain molecule, the primary food for your colonocytes (colon lining cells), which microbes make from digesting starch, but can be in high proportion of some ingested fats like coconut oil. You will hear much about the health-giving benefits of coconut oil therefore but, like everything, it is not a panacea for a badly balanced diet—you must be ingesting enough leafy vegetables for it to be beneficial. We also need more prebiotic fiber such as inulin (found in onions, garlic, dandelions, Jerusalem artichokes, etc.). This encourages the right bacteria to grow in our guts. Prebiotics are the foods for the probiotic bacteria, and you need to provide both or the changes in gut flora won't last. If you compare how much fiber we ate as primitive peoples (and even up to the Victorian age), it becomes obvious that we have a much-reduced fiber intake. This directly affects the diversity of our gut flora, an interaction that has become known as the "keystone relationship"[Velasquez15].

A healthy diet should include low omega-6 fats (we have enough), high omega-3 oils such as hemp seed, flax seed, and natural unprocessed oils such as olive oil and coconut oil. To improve the natural fats quotient, add nuts and seeds into your diet including nut butters. On the subject of butter, it is a good fat (provided it is organic to avoid fat soluble pesticides) as it contains butyrate and fat-soluble vitamins (A, D, E and K). Avoid sweet fats though (e.g., ice cream); substitute those with frozen smoothies using natural fats such as avocado and coconut.

Prebiotics and Fiber

Fiber is a form of slow-release carbohydrate which is the favorite food of good bacteria; they take fiber and turn it into butyric acid, and that helps us have energy. But, like most nutrients, fiber is not all the same and only specific types are beneficial.

Galacto-oligosaccharides (GOS—present in breast milk and some complex carbs) are better than *fructo-oligosaccharides* (FOS—found in simple carbs, for example, glutamine and inulin) as they don't cause bloating, so are useful in treating conditions such as *small intestinal bacterial overgrowth* (SIBO[2]). This is a condition whereby the bacteria

---

[2.] SIBO is a common condition. L plantarum is a good treatment. Oil of oregano and berberine can also help but you might have to take low-dose antibiotics if it is really bad.

normally found in the large intestinal (colonic) have migrated up into the small intestine, where they're not meant to be. The small intestine (ileum) is designed to have a different collection of bacteria to the colon —and a lot less! If you have the wrong bacteria there, you get a lot of bloating and gas lower down the system. A sure sign of SIBO (apart from bloating), is sensitivity to fermented foods, for example, wine and beer which will trigger you. You will also not be able to take the fermented tea Kombucha (which has wild yeast in it). "FOS, and inulin particularly, can also feed yeast like *Candida*, so people with active yeast infections should also limit their intake" [Brisson20] as the body is on high alert for yeast and will react quickly with a higher histamine reaction, for example, flushing of the face after imbibing these. Histamine causes this flushing of the skin by triggering inflammation in the body. As I've already described, inflammation is a normal part of healing which dilates blood vessels in the area to allow healing agents to flow there in the blood. But it can become chronic, and certain foods can trigger this. You will have to experiment with different foods to limit FOS.[3] And if you are getting flushed after certain foods, then suspect SIBO as the cause.

Complex carbohydrates also work by supporting growth of Bifidobacterium and stimulating the immune system. Other good fibers are psyllium or flax seed (ground in a blender) as they liberate more lignin— this is a phyto-estrogenic substance (an estrogen receptor modulator derived from plants) which helps regulate hormones. High estrogen is a real problem in our environment both from the waste water supply (high amounts from the birth control pill is excreted in urine) and xenoestrogens (false estrogens) in the food supply from pesticides and plastic packaging. When the gut bacteria transform these phyto-estrogens, they bind to hormone receptors and help to block out endogenous estrogen (specifically estradiol—one of the three estrogens). This binding is good for human breast and prostate cancers, which are known to be linked to high estrogen levels, incorporation of more alkaline foods (vegetables generally) fosters a better balance of microorganisms.

You can also learn how to garden; grow your own or buy organic. Make sure you inoculate yourself with microbes in dirt—they make you happier

---

[3.] Interestingly, the FODMAPS diet was adopted by the NHS as a solution for people with IBS as it limits complex oligosaccharides and replaces them with simple carbs (some would say junk food). This is no long-term solution; it limits the food without treating the overgrowth of SIBO and yeast, which is the real problem. However as short-term relief it can be useful. But I would rather treat with probiotic bacteria and antifungal herbs.

due to containing microbes that help make "happy neurotransmitters" like serotonin.[4] Therefore, take up gardening—even if it's only to plant up tubs. Don't scrub your organic produce, eat more dirt, and cook from scratch if you can.

Natural Probiotics—Fermented Foods

Apart from lack of fiber, the other lack in our diets compared to ancient cultures, are fermented foods, which are already enriched with good bacteria and the enzymes they produce that help us to digest. Fermented foods are common in many cultures especially eastern European, but fast food, refrigeration, and the decline of home cooking have seen a slow reduction in preservation and culturing of fermented foods (beer doesn't count). We will cover this in more detail in the next chapter as it is an important way in which you can improve your microbiome at home for very little money.

Nutritional Solutions; Special Diets

There are so many dietary fads that come and go. The latest at the time of writing seems to be "clean food," which is interpreted variously as gluten-free/dairy free and or alkaline (depending on who you ask).[5] The problem is some of these ideas get taken to extremes by certain people with a product or lifestyle to promote. And, because so much of eating is psychological, we have to take into account that it will play into the hands of those with a disordered relationship to eating (anorexics, etc.). The mix of self-improvement and addiction to the high of "the cleanse" is very potent. Indeed, there are always fad diets every year (January seems to contain most of them).

Part of the problem is that the message put across by all these diets and fads gets simplified and "dumbed down" to make it media friendly and the *complexity* of approach, that I am trying to promote here, is lost. We then get a period of excitement and media frenzy followed by the inevitable backlash. Much of it is formulated in terms of "pseudo-science" versus "real" science. Again, this is a polarized approach that does nothing to help promote people's ability to get motivated to *do* something. If they are so confused, they can't work out what's good for them or not, they tend to do nothing.

[4.] See Mark Sisson's book The Primal Blueprint and website for more information on this http://www.marksdailyapple.com/eating-earth-exploring-the-mysterious-world-of-geophagy/

[5.] The BBC Horizon program "Clean Eating" was aired as I was writing. www.bbc.co.uk/programmes/b08bhd29

I personally liked the natural approach promoted by the authors of a recent book on fixing their father's diabetes by changing his diet.[6] Many reviewers expressed utter disbelief that this was possible, despite the evidence that it had been successful. There has been a recent spate of critical articles, a book, and even a blog by the self-styled "Angry Chef" on debunking food fads who blames the gullibility of the public to diet gurus [Lewis17]. I don't want to get into that particular argument here, as I actually find myself somewhat in the middle—neither do I agree with them completely or with the clean food movement that tends to attract a largely affluent, middle-class audience. In some instances, with online blogs and other media, clean eating can be deliberately targeted to anxious people (especially women) who can get sucked in to restricting their food even more than they already are, that is, it can play into food obsession/anorexia.

In any case, I think as a basic given we could all do with eating more vegetables for the reason of both remineralizing and alkalizing the body. In addition, trans-fats (chemically altered) and excessive saturated fats (particularly those in baked goods) have a negative effect on gut bacteria so should be limited. Remember, it's not the amount but the *type* of fat that is important. Those are two things everyone can agree on and it's relatively easy to do to make a difference. So, I will consider just a few highly research-supported diets here.

Gluten-Free Diet

Many people are now switching to gluten-free foods (GFF) either because they have celiac disease and cannot tolerate any gluten or, more commonly, they have discovered they are healthier without gluten in their diet; they are more properly termed *gluten intolerant*. However, gluten-free foods are not necessarily better for you unless they are *naturally* gluten free. The sort of foods often sold in supermarkets as GFF are often highly processed and have more added sugar to give flavor. They are *not* healthier alternatives but just junk food by another name.

If you find that you are gluten intolerant, then it is wiser (albeit less convenient perhaps) to substitute foods that are naturally without gluten. This requires a lifestyle change rather than a simple "diet." But many people who do this report improvements in their energy levels, immunity, and sleep. As the idea of being gluten intolerant (i.e., not pure celiac but having *nonceliac gluten sensitivity* (NCGS) is still highly controversial, giving

---

[6] http://www.fixingdad.com/

up gluten is not something you will be encouraged to do by your doctor. However, this is slowly changing as more and more evidence emerges of rising levels of NCGS.

I certainly feel better with low or no gluten in my diet. One of my first symptoms was the headache, fatigue, and bloating (a gut-mediated immune reaction) to pasta; now that I have corn or rice pasta, those reactions have largely disappeared.[7] These alternatives are readily available in supermarkets at two or three times the price of their nongluten-free cousins; that and gluten-free bread have seen a phenomenal rise in sales which shows how much of a problem it is for a large section of the population. This is generally the case with most processed foods; to keep them strictly gluten-free means the processing is that much harder and the low numbers compared to gluten-based foods means the economics are still disproportionately less favorable to the consumer.

A gluten-free diet is still the first-line basis of many naturopathic (natural medicine) treatments, and I would concur that, for the majority of people with chronic illness, it is a major self-help intervention that can make a big difference while you are dealing with some of the symptoms. Whether it needs to be a life-long restriction is hotly debated.[8]

Calorie Restriction and Intermittent Fasting

The truth is it's not just what you eat but *how and how often* you eat it. One of the most important modifiers of your health, is simple caloric restriction. This fact created quite a stir when research on rats began to show those who ate the least in terms of calories, lived the longest. For a while it spurned a whole movement of people dedicated to living on a low-calorie diet *forever.* They certainly showed very marked positive changes in biochemical markers like blood pressure, blood sugar and so on. But having seen some of the restrictions that need to be put in place and considering the differences between rats and humans, it occurs to me that this simple *denial in pleasure from eating* (as most of the calorie restriction seemed to do) may have more to do with a punishment ethic. Indeed, I think there are other ways you can restrict intake that are friendlier to yourself: the

---

[7.] I did, however, notice that they would reappear if I didn't take additional digestive enzymes to help me digest the starch. Sometimes the issues change over time as the gut gets more and more leaky. This is a sign that you are not dealing with the root cause.
[8.] Certainly, this is the current advice with CD or NGGS. But there is a compelling argument that the gut can be healed and thus the restriction lifted.

5+2 diet which modifies eating to a lower calorie intake on two days out of seven seems to be a more pleasurable option and therefore one that is easier for the majority of people to achieve. Valter Longo has been the main proponent of this approach with his Fasting Mimicking Diet [Longo22] where you restrict your intake to a low 500–1000 calories per day for five days.[9] But even more basic and appealing, is the "fasting window"; the idea is that you make sure you have at least nine hours (preferably twelve as in Longo's Longevity Diet [Longo18]) between the last thing you ate at night and the first thing you eat in the morning. This gives your body a chance to repair and clear out the remains of the last meal you ate—it promotes cell renewal and repair or "autophagy." If you snack in the evening this important mechanism is overridden, and you will be building tissue instead of breaking it down.

Ketogenic Diet

Another idea that is currently gaining ground is the ketogenic diet. This is basically where you restrict carbohydrates severely, moderate protein, and increase fats—very much like the Atkins Diet (except Atkins did not restrict protein). When you cut the carbs in this way, the body is forced into *ketosis* where the body runs out of fuel and is forced to use fats for energy. The body then becomes more efficient at burning them and converts them to ketones in the liver; these are then transported to the brain. This will happen inadvertently when you are ill and stop eating for a day; it is a normal process that allows people to direct energy to recovery rather than digestion. Some people report feeling clearer in their heads during that time largely because the brain is forced to use ketones too and they are less inflammatory molecules.

You will need to fast for at least twelve hours to get ketones to form; thirty-six hours is normally considered average for full ketosis. Entry into ketosis is "normally marked by a characteristic 'sweet' breath odor (..) caused by acetone, which, being a very volatile compound, is eliminated mainly via respiration in the lungs" [Paoli13]. Ketogenic diets have been a popular fad for a while now, and they have been found to have various health benefits including lowering blood cholesterol and encouraging weight loss without calorie counting. There have been some studies showing reduction of risk and even improvement for Alzheimer's [Hertz15], epilepsy [Martin16],[10]

[9.] See https://www.valterlongo.com/fasting-mimicking-program-and-longevity.

[10.] This is, in fact, how ketogenic diets were first discovered as a "cure" for epilepsy in the 1920s.

heart disease [Nordmann06], and even cancer [Erickson17]. Clearly it has much to commend it – particularly as a "neuro-protective" diet. On a practical basis, it does seem to suit some people very well, but others struggle more, perhaps because of differences in fat metabolism. Ideally, of course, we should also be sure to nurture the gut as a whole rather than just restricting one nutrient. In any case, a low carb diet, rather than low fat, is the most important change.

MCT Oil

A variant on the ketogenic diet is adding medium chain triglyceride (MCT) oil to your diet. There are a variety of products so marketed which go under bizarre names like "brain octane oil" (coconut oil extracted for MCT). MCT contains medium-length carbon-chain fats like coconut oil, but it's not clear whether it has any special advantages over coconut oil itself, even though many such claims are made. Some people seem to tolerate it better.

One way of adding it is to make your coffee blended with coconut/MCT oil instead of milk as an afternoon pick-me-up. I have tried this—it's quite fatty tasting but does seem to do the trick.[11] A high-fat MCT diet adjusts the microbiome; the good bacteria like Bacteroidetes grow preferentially. When we eat colored fruit or coffee containing polyphenol antioxidants, we are also preferentially feeding the "good" bacteria. So, in this sense, coffee acts as a prebiotic. You have to weigh up the pros and cons. I drink coffee, but I limit my amount to one or two cups a day, and I make sure it is always fresh and good quality (ideally organic).[12] You have to experiment and try what works for you; if you are sensitive to caffeine[13] then you may be better off without, or at least having days without. There is no "one size fits all" solution in my view with any diet.

Rather than list all the various food fads, I am going to consider some simple procedures and techniques that everyone can add to their lifestyle. Finally, I consider certain protocols that have worked in highly specific situations.

[11.] However, beware some coffee grounds, and particularly cheap instant coffee, may contain mold (mycotoxins), so you may get chronic low doses from coffee. These have a detrimental effect on the microbiome as they promote the formation of biofilms that make it difficult to eradicate the yeast. Therefore, limit the amount.

[12.] A good place to get organic coffee if you can't get it locally is at IKEA. They have made sustainability and fair trade a key part of their ethos.

[13.] This may depend on your particular biochemical individuality, specifically your neurotransmitter dominance. See The Edge Effect Quiz by Eric Braverman on https://www.bravermantest.com/.

Body Ecology Diet

We must stop using the terminology of war in our relationship with health and healing, that is, instead of looking at killing the "bad bugs" we look to establish more "body ecology." With our growing understanding of ourselves as a community, the idea of ecology becomes ever more relevant.

There is a fantastic protocol developed in the United States by Donna Gates called the Body Ecology Diet. In many ways, it is similar to the GAPS protocol, and it was developed to solve the same set of problems. It differs in some of the specifics but overall, it is about balance and harmony in the digestive tract. It is a plant-based organic food diet, with an emphasis on fermented foods and eliminating sugar and gluten but in a way that feels more about pleasure than restriction. A real pioneer in the field, Donna truly embodies the diet she promotes as it's more a lifestyle change than a "fad" diet. I highly recommend you check out her website for recipes and ideas [Gates11].

Gut and Psychology Syndrome: GAPS Diet

Developed by UK-based GP Natasha McBride, the Gut and Psychology Syndrome (GAPS) diet is designed to heal the gut and therefore prevent the absorption of toxins. It is more of a protocol not a "diet" as commonly understood. It is based on the belief that all diseases begin in the gut. She claims to have had great success rebuilding the gut and reducing neuropsychological symptoms, for example, in autism.

Unlike some of the other nutritional interventions covered here, this one goes further than most in removing all foods that are difficult to digest and provoke a reaction, to leave healing foods. This includes grains, (even gluten free—starch is a large molecule and is difficult to digest), cellulose (fiber is difficult to digest and may go through you and feed pathogenic microbes and, land in your bowel).[14] If you have healthy gut flora they will then digest the fiber and starch, but if your digestive system is dominated by pathogens it is the pathogenic bacteria that will grow larger and create illness in your body; they all love fiber. Here is the basic protocol:

- Remove all complex carbs (including parsnips, potatoes) for two years.

[14.] Our bowel is our vegetarian gut; like the herbivores rumen where bacteria digest the grass for the cow to short chain fatty acids (SCFA—the building blocks of fats). Our bacteria should help us do similar things with fiber.

- Remove all dairy unless fermented (e.g., yogurt and kefir—both are predigested and so easier for the digestive system to handle[15]).

- Add fermented vegetables, fish, and most beans (green beans and navy beans can be introduced later in the diet, but they must be soaked). Avoid short-chain oligosaccharides (the building blocks of carbohydrates).

- Vegetables are good—microbiome friendly, including radishes,

- Limit sugars as they attach to hemoglobin glycosylating them (making them "sticky") and this may block capillaries causing amyloid plaques leading to blindness, Alzheimer's, and possibly metabolic syndrome.

The protocol starts very strictly and gradually more foods are introduced: soup for breakfast, lunch, and dinner—meat stock (not bone broth) provides the "bricks and mortar" for healing and rebuilding, so this comes first. We need to eat meat for its essential fats, protein, and gelatinous tissue. It is good to ingest meats such as the inner organs and bones, tongue, tripe, liver, heart, and skin (but not muscle as this is the most difficult to digest). This is cooked slowly to enhance glucosamine content as collagen is released (such as in a slow cooker or over a natural fire). The protocol continues after that with:

- Fermented foods and well-cooked nonfibrous vegetables, leeks, onions, cabbages, cauliflower, and courgettes (zucchini) until the symptoms vanish.

- Eggs, more vegetables, more fermented vegetables, supplements to bare minimum here. Fish oils, multi amino acids, and some probiotics (high diversity). Some people have other needs like glutamine. Meat stock will provide most of this but in a balanced form so better than supplements.

- More complex foods with high animal fat: pork, goose, duck fat, butter, and ghee.

Natasha cautions that one may need to move back and forwards as you see symptoms reappear. Each person will have a tailored program to take into account their constitutional strengths and weaknesses. Some people have a damaged gut, but the gut compensates, so they have no digestive symptoms. Instead, it is the kidneys or the lungs that show the problem, for instance. Most people recover without supplements, she claims.

[15.] Note that raw milk may be easier, as it has different bacteria present that help to digest, but this is not mentioned in this diet.

She believes we need to use our senses to rely on what our body needs; remember metabolism is dynamic. Babies smell, touch, and smear/taste their food and their body gives feedback. Parents should not stifle their children in this area as it's important to develop a normal relationship with food, she cautions. Equally as adults, we need to redevelop this ability. I can't disagree with her on this. My only concern is that you would have large noncompliance if you tried to get the majority of people to take on this diet. So, although I am sure it works (and indeed some of the principles are used elsewhere very successfully), you would have to be quite ill, in my opinion, for this to be seen as an attractive option.

Eat Fat Get Thin

There is no major controversy more heated than the controversy around eating fat; it continues to stir up passions in the world of nutrition. For many years, we have been exhorted to eat less saturated fat and increase our polyunsaturated fat intake. However, this has often been at the expense of our health; people who cut fat out of their diet are often less satiated (full feeling) and thus will overeat carbs as compensation. It is not avoidance of fat that keeps us thin but eating *good quality fats*. Natural fats that are derived from organically reared animals may be saturated but are of good quality. Vegetable oils that have been cold pressed, so they have the color and flavor of the originating source (e.g., olive oil) are also good. Stay away from artificially produced fats like vegetable or canola oil. Moreover, particularly avoid skimmed milk which is not food at all, and bears little resemblance to natural milk. If you want to lose weight eat raw milk and butter, but avoid margarines, which are very bad for you. They are highly inflammatory and processed; some actively destroy health by containing "high levels of trans fats which packed a double whammy for heart disease by raising levels of LDL (bad cholesterol) and lowering levels of HDL (good cholesterol)" [HHP20]. The idea that they somehow help protect your heart has been shown to be based on flawed data.

Food Preparation Traditions

Blending

I recommend every one of my clients get a blender (best with a shaker cup for on the go), and also buy some protein powder for adding to the juice/smoothie for extra vitality (e.g., for breakfast in a hurry). There is a vast difference between a blended juice and a store-bought juice (they are really sugary). For instance, making your own cucumber, celery, kale, parsley, lemon wedges, and apple blend is really good and tasty as long as you get the proportions right. Blending has the advantage that you do not waste any of the material, but you have to make sure you follow certain recipes that work. A few times I've had to throw away a promising blend that turned out undrinkable as the consistency was too dense.

Juicing

Juicing is similar to blending but you buy a slightly more expensive piece of kit and you waste some of the fiber as the juice is extracted from the skins and fibrous material that makes the resultant drink more palatable. I have both a juicer and a blender for different times of day but if you only want to purchase one, I'd go for a blender as it is multifunctional and generally takes less cleaning (one of the downsides of juicing).

Figure 4.1. Compact juicer.

There was a tremendous fad to juicing which took off about ten years ago—suddenly there were juicers everywhere. You will most likely find them redundant and unloved in thrift stores now. The issue for most people is that it takes a lot of preparation with chopping and cleaning up. Still, if you have the space to store one, I'd recommend having your juicer/blender on the work counter rather than in a cupboard. And modern juicers are much simpler machines that take less dismantling when cleaning.

Slow Cooking

Buying a slow cooker emulates older styles of cooking whereby food would be cooked at low temperatures (sometimes all day) in the embers of the fire or, in some cultures, where it is available using geothermal energy (burying a crock pot in the earth). Meat becomes much more tender that way, and vegetables stew in their own juices too, making the resulting food more nutritious than boiled or fried food. It is also a way to get the goodness out of meat bones, whereby the collagen from the bones is broken down to be made available for your body.

If you have more time in the morning/lunchtime to do your preparation than you do toward the end of the day, or you run out of energy to cook, this could be a good solution.

Don't forget that stewed apples (also termed applesauce in the United States) are a form of slow cooked fruit that has huge benefits. If your diced apples are heated gently in a pan with water covering approximately one third of the height of the fruit with a pinch of cinnamon, until after roughly five minutes the skin develops a sheen, you know you are done. This sheen is due to the release of pectin from the apples, a compound that contains an important anti-inflammatory chemical called *intestinal alkaline phosphatase* (IAP), that protects your body via:

- binding bacterial LPS (an endotoxin) and escorting it out with the stool

- stimulating the genes to heal leaky gut (gut permeability)

- stimulating good bacteria to colonize and rebuild

Fermented Foods

Fermented foods are an absolute essential to well-being—they allow you to eat foods that are partially predigested by their own bacteria. Given that, as we have seen, we have what has been termed "mass dysbiosis'" in Western nations, most people have imbalance in their gut flora causing all manner of preventable diseases, it behooves us to do something to help us restore our gut flora. All diseases are a disconnection from a healthy gut.

Fermented foods are foods that been allowed to be acted upon by digestive bacteria. A common example is yogurt. I remember when yogurt was a rare food in the UK. It came in one flavor (Ski) and was exotic and strange … now there are so many different types it's bewildering. But there are other types of fermented foods like sauerkraut, kimchi, kombucha tea, and so on—they all have exotic sounding names because generally each culture has their own variation. What they all share is that they are incredibly high in probiotic bacteria; 100 trillion probiotics for 2–3 grams of fermented foods. Therefore, eating fermented foods is an example of natural supplementation without having to buy expensive supplements.

Not only is it a natural mixture but it is cheap to make usually once you learn how. Some people with high fungal infestation may need to avoid it until they've healed the leaky gut and fungus first. Otherwise, it can cause an extreme reaction—I can't have kombucha for instance but can handle yogurt and kefir (especially the raw/nondairy types).

The gut microbiome not only helps you digest food, but it adds vital vitamins and minerals to your body, for example, vitamin K2 (which works synergistically with vitamin D helping to take calcium out of your blood vessels and puts it back into your bones where they belong). There is currently no standard test for K2 so is not often known or tested for, but it is a vitamin that helps coagulate your blood as well as help you maintain a healthy heart.

Every bit of understanding you've received so far in your life has been about "germs" and how to eradicate them. So, at first it can be a bit scary, but you start to trust and tune into your senses more and your intuition. Your ancestors used their gut instinct in this way. Animals have this—they sniff food really well before committing to eating it—especially if it is new. We have largely lost that sense,[16] but we can bring it back when we understand its importance as the pillar of our healing. There are many principles of nature rooted in our gut instinct. So, by getting in touch with our bodies again changes many areas of your life. Dr. Raphael Kellman of the Kellman Wellness Center encourages us to "get befriended by nature" again, which is great way of expressing this urge.

Benefits of Fermenting

The benefits of fermenting are many and varied. Some I have already touched upon in terms of long-term health. However, you may be more interested in the immediate effects, some of which are:

- better digestion

- more energy

- healthier skin

- weight loss (natural)

According to clinician Noah de Koyer, "healing is an inside out job," which in my opinion also, is absolutely accurate. Let's now look at the process of fermenting or culturing.

[16.] And chemical foodstuffs deliberately manipulate it, so it's difficult to tell.

The Fermenting/Culturing Process

Learning to ferment at home is a good move to nurture your gut. And these days it's not such a difficult thing to do; you can take a course either online,[17] or if you have a group near you, in person. Most people start with sauerkraut as it's relatively easy to achieve. The basic idea is that you chop your veggies, add salt and let it ferment. All food should naturally contain bacteria—they are not universally to be feared. We have gotten so used to hearing about bacteria in food as a threat, it comes as a shock to learn that in some circumstances they can actually be *beneficial*.

Fermenting is an art form (like cooking) and takes time to master well (and I can absolutely attest to having made my fair number of mistakes). We are going to look at the four main types of fermented foods now in the following section.

Functional Ferments

Functional ferments include sauerkraut, kimchi, yogurt, and kefir, and contain a wide range of probiotics. Don't bother buying in the ready-made pasteurized probiotic that has been heat-treated and therefore contains little of value; it needs to be raw if you buy it in (and that's difficult to find as it goes off quickly). It is much better to start making your own if you can. Luckily raw milk will naturally ferment if you put a little aside and let it go "off." Even some beer (old style dark beer or ale) can be good for you—monks used to use it for health and ordinary people drank it as it contained minerals). This is because the grains are predigested, which helps our weak digestive tracts by harvesting the power of plants and bacteria to help us digest. There are other less well known fermented foods such as Elis (lentils and rice), and Engira (Ethiopian edible spoon), so you can see the range is very wide in different cultures. There is a good research base on the health-giving properties of these foods [Prado15]. But first, I am going to look at the most popular that you will find in the western world: sauerkraut.

Sauerkraut

Sauerkraut comes from the German for sour cabbage, from which it is traditionally made. However, there are many variants which contain other vegetables for increased benefits and flavor. To make standard sauerkraut is remarkably simple:

- First, chop your cabbage in a bowl.

[17.] Summer Bock runs a particularly good one called "Guts and Glory.".

- Massage 2–3 teaspoons salt in (salt is really key to the process, so do not leave it out, but you need good quality Himalayan/Celtic salt—not table salt—which is highly processed and doesn't contain any minerals).

- Add a little water in a Mason jar, so all the cabbage is submerged. Use good quality water (not tap water if you can avoid it). Leave on the side with a lid (and bubble collector). You should start to see oxygen bubbles within an hour or so, which gradually speeds up, much like brewing beer.

Figure 4.2. Sauerkraut fermenting.

Sauerkraut is usually done in 5–10 days depending on the outside temperature. The next step is to swap the lid for a sealed one and move it to the fridge. I have a jar in there from last year, but I would not recommend keeping it that long as the ferment starts to get very strong!

You can modify the culture by using a different starter. For instance, Lactobacillus plantarum added to the fermented vegetables as a starter seems to increase the probiotic content 3.5 times. Use a starter culture such as Kinetic Kulture from drmercola.com. It is high in bacteria that promote vitamin K2. Even if you can't get this one, start anyway, don't worry about getting it right first time. Everyone will be different so it's a personal exploration.

The results will be initially better digestion, clearer skin and eventually you may notice weight loss, sometimes quite substantial. Obese people have a different balance of gut flora of Bacteroidetes and Firmicutes as we've seen in Chapter 3. Most diets fail because this is never addressed (alongside the psychology of dieting promotes scarcity and therefore fat retention). Different diseases/conditions might need different types of culture; this hasn't been fully explored yet. The acids and enzymes that the right bacteria provide are good for you when they are in balance. They reduce inflammation (which is the cause of most disease and lower "bad cholesterol" production (i.e., the protein-lipid complex which it refers to).[18] The easiest and cheapest way to do this is by using fermented foods.

A big proviso though: start slowly by adding 2–3 spoons a day to your diet. Less is more here, and this is plenty to begin. You need to get your gut used to this; after all you are adding some foreign bacteria and it will take time to adjust. Some people have made themselves very sick by eating too much in one go too soon.

Kefir

Kefir is like yogurt, both in taste and process, but it is a little bit runnier and can be drunk rather than eaten. It is beginning to be quite popular due to its convenience (easier to add to a busy lifestyle than sauerkraut). It can be added to a smoothie or your breakfast porridge/muesli. Once you have the grains (which are the organism that makes it for you—they look like white frogspawn), you can keep the production going very much like with yogurt. However, two things to be aware of are to keep your jars sterilized before use (or you will contaminate with less friendly bacteria), and to not use metal spoons or implements when straining and manipulating the stuff. But easier by far is to begin by purchasing a starter culture (dried bacteria). Different starters change the nature of the eventual ferment, but for a beginner they make it more failsafe than using grains.

The same companies that provide information and cultures for sauerkraut often produce kefir as well. You can now get the ready-made product delivered to your door too via such wonderful businesses as Riverford Farm and Chuckling Goat (UK only, no doubt there are many other options in the United States and other countries). It has become something of a "cottage industry" and many people are finding this is a good alternative to drugs for such skin conditions as eczema, psoriasis, and so on.

[18.] There is much confusion in the popular press as to what cholesterol is and what it does. Cholesterol itself is a natural molecule needed for the production of hormones. The cholesterol/protein complex commonly termed good or bad cholesterol is what the body uses to "seal" the damage done to blood vessels by inflammation.

There is even a range of cleansers and lotions with kefir added that can be applied topically. Many of my friends have tried this and had great results. I have tried to ferment my own with mixed success, for example, leaving the grains in the fridge too long in the finished ferment is the most common mistake. They run out of food and die. Then, you just get cheese or slightly rancid milk. For best results, you need to strain your mixture after twenty-four hours at room temperature to remove the grains (omit if using powdered starter); in which case strain off about a third of the mixture and add back to a fresh batch of milk. You can leave the ferment in the fridge if you don't use it immediately but too long and you kill the grains as I have found. Getting a good process flowing where you make just enough for your own use and re-use in a regular routine is ideal. There are plenty of instructions (and videos) on the internet to show you how. I reiterate; do not use metal spoons or bowls, keep to wood, plastic, and ceramic implements.

Fermented Soy Products—Tofu

For increased assimilation, you can't beat fermented soy products: tofu, natto, and tempeh. By fermenting the soy, it seems to allow people to tolerate it better if they are normally sensitive to soy. I would not recommend unfermented soy milk, for instance, as a general rule—it's okay in small amounts but best to add other milks to your diet too, as soymilk has compounds in it that can encourage hypothyroidism if you do not have enough iodine in your diet (which is common). These fermented soy products are generally bought in ready-made. I have not come across anyone (yet) who has made their own. But they are readily available in good whole food shops and online. Be aware that, although natto is particularly high in vitamin K2 and encouraged therefore, its consistency often puts many people off.[19]

Fermenting for Preservation

Some foods are fermented to preserve them—this was useful in the past when there were no refrigerators. For example: vinegar, cheese and pickles. Coffee and chocolate are also in a sense fermented foods, and so can be considered health foods when the coffee is fresh and the chocolate dark due to the polyphenols they contain. However, as there are no microorganisms present, I am ignoring them here.

[19.] It's hard to describe what natto looks like without being disgusting. I'll leave it for you to check out online.

Kombucha Tea

Kombucha is a fermented tea product made by a fungus (yeast) "starter" in caffeinated tea. You can add various natural flavorings to make the tea more palatable. It is created by fermenting with a fungus called the "mother" or "SCOBY" (this stands for symbiotic culture of bacteria and yeast). Of course, these commensal (mutually beneficial) associations exist throughout nature. The fungus can convert the caffeine in the tea into carbon dioxide, and so the resulting ferment has a slightly fizzy taste. I have had limited success with kombucha, although I have tasted some lovely varieties made by other people. I think my technique is just not good and I end up with a cloudy solution with, presumably some mold spores still in it. It can trigger symptoms somewhat like candida overgrowth, therefore. Some people claim amazing health benefits. However, it may be more of a problem if you have a high yeast population in your gut which is already producing some toxic by-products, to add more yeast to your system (and if your detox systems are compromised). Certainly, this has been the case for me, although I know many people who tolerate it very well and it forms a great substitute for carbonated drinks (the yeast forms bubbles in the drink), particularly in the summer months. However, it has many beneficial constituents when made properly. I would recommend using a specialist supplier of starter cultures[20] if you are interested in trying this at home.

Psychoactive Ferments

Alcoholic drinks like beer and wine are technically psychoactive ferments which used to be considered a health food when they were home produced. But commercial yeast and stabilizing/preserving chemicals make them more likely to produce an immune reaction. If I drink wine or beer, I try to make it an organic variety. I will not consider them further here.

Technically, kombucha tea falls into this category as it has caffeine in it before it is fermented, and it can produce alcohol so it is not good for people who are sensitive to alcohol or avoiding it for health or addiction reasons. The amounts are relatively low, and in a good ferment negligible. More concerning for me is the amount of mold that may build up in the liquid.

---

[20] In the UK I have found Happy Kombucha to be excellent in this respect; selling everything you need to ferment your own.

Probiotics

We have discussed probiotics already with regard to detoxification. Here, I am referring to artificial collections of bacteria which people take either as a result of cleansing or as a natural vaccine to help repopulate the gut flora and therefore regulate the immune system. The best are natural foods; it is true that pill form supplements don't implant well and alter the terrain long-term. Sauerkraut and other natural probiotics wash through and inoculate the system better. But there are times when the digestive system has been so depleted that some additional support with supplements is necessary. Signs that you may need them might be that you have digestive problems (constipation or diarrhea). For instance, there has been much research into this with colitis, atopy (allergy), irritable bowel syndrome (IBS), and so on; probiotics have been shown to be really effective [Pace15].[21] I have experience in treating my own IBS with low-dose probiotics after a forced bout of antibiotics, and it made a difference very quickly. The effect is immune-regulatory in nature; it changes the expression of the messenger molecules within the gut lumen (inside) and therefore the coordination of immunity throughout the body.

Two words of caution though with probiotics: more is not necessarily better. You need to be careful with high-dose probiotics. Start with low doses (5–10 billion CIU) and with multiple strains, at least at the beginning. This is especially true if you have high GI symptoms particularly like SIBO, and so on. You need to go gently to balance the microbiome by changing the terrain. You also need to rotate probiotics because you don't want overgrowth of particular strains. I would encourage you to try different types and to remember to eat more soil, as it contains soil derived Bacillus types.[22]

Particular strains have uses in particular diseases; for instance, Lactobacillus plantarum (another soil-based bacterium) tends to help rebalance overactive Th1 (innate immune) activity that often manifests as frequent low-grade infections. It boosts immune function, and so helps everything from preventing wound infections to lowering anxiety and depression. Bifidobacterium is another important one in brain function. In summary, there is no one probiotic that is best. Try a balanced formulation by a reputable company and notice how it affects your function and change if it doesn't work. Monitor biofeedback from your own body. Better still: get advice from a professional who can tailor a protocol to you.

[21.] See greenmedinfo.com for fully indexed studies.
[22]. That is, don't scrub your organic vegetables and eat more of them raw.

Second, to paraphrase researcher Shaheen Lakhan, restoring a sick microbiome is not as simple as adding back missing or unrepresented species. Both the physical habitat of the gut and nutrient resource delivered must be durably changed in order to produce a durable change in the microbiome [Lakhan10]. So, it is more about *making the terrain hospitable* than making short-term changes in personnel. Get the environment right first.

Phyto (Plant-Derived) Nutrients

Finally, we come to one of my favorite subjects; modifying your function with edible plant-derived nutrients (phyto-nutrients) like chlorophyll, resveratrol (in red wine and grapes), and vitamin A (retinol). I talked about these in more detail in my last book in terms of their similarities as *chromophores*. This is a term used to describe a natural compound that is high in double bonds between the atoms that resonate (vibrate) at high frequencies in the body and absorb and reflect light; they are responsible for the bright color of the food.

The electron transport chain (as described in Chapter 2) in mitochondria contains many different chromophores that are integral to the release of energy within that system. Up until now we have always thought that it was the unstable electrons that these molecules possess that release energy, but it may be that they harvest light energy via photons (the smallest particle of light) too [Xu14]. Whatever the mechanism, chromophores are now understood to be very important molecules in terms of the microbiome, as bacteria use them selectively as substrates to convert into vital biomolecules.

Chromophores and the Electron Transport Chain

For instance, recent research into chlorophyll (the plant pigment that makes leaves green and is able to photosynthesize sugars from sunlight), has shown us that gut bacteria are able to convert it into a molecule[23] which, when taken into mammalian mitochondria (site of energy production) changes one of the molecules in the electron transport chain (ETC) to the reduced form. This increases production of energy *without increasing oxidative stress of free radical formation*. Chlorophyll thus makes energy production more efficient and less damaging to the body.

[23.] Called pyro-pheophorbide-A, which reduces ubiquinol in the ETC—omitted due to horrible chemical terminology.

This important conversion of chlorophyll enables biologists to reclassify our species to a photo-heterotroph (or autotroph) that is, *we can directly convert sunlight energy much as plants do.*[24] This information will revolutionize our understanding of our biology. It is very new information so don't expect to find it in any textbooks yet. The details remain to be discovered.

We have often been exhorted to eat more antioxidants in order to quench the free radicals formed as part of the ETC and they are understood as being important regulatory molecules in human metabolism. But the story is, as usual, more complex and more diverse than that. There is a need for plant compounds with chromatic properties called bioflavonoids (or polyphenols—both terms are used synonymously). The mitochondrial signaling of the ETC responds to the full color spectrum of foods (each complex in the chain has its own resonance frequency) and thus a diet rich in variously colored foods is essential. These molecules give food its color as found in various fruits and vegetables. Fruit is particularly rich in antioxidant molecules called bioflavonoids, for example, blueberries, strawberries, and grapes—are particularly potent sources of bioflavanoids like ficetin which protects regulatory T-cells (T-regs), part of our immune system.

Vitamin A and the Carotenoids

We know that brightly colored vegetables are good for us, because they contain bioflavanoids called carotenoids (a precursor of vitamin A). Vitamin A (retinol) is the active compound formed from beta-carotene in orange-red vegetables like carrots. Retinol and its metabolites have been found in the vertebrate eye, so we know they are important sensing molecules, but recent research shows us that it has a "hitherto unsuspected role in mitochondrial bioenergetics, acting as a nutrient sensor. Retinol is essential for the metabolic fitness of mitochondria" [Acin-Perez10]. Interestingly, eating a high-lipid-containing food like avocado with your carrots enhances absorption of alpha- and beta-carotene and, more importantly, conversion to vitamin A [Kopec10]. So, add olive oil to your salads for extra absorption.

ATP: More Than an Energy Molecule

According to Dr. Heinrich Kremer and his cell synthesis theory, ATP serves as an 'antennae molecule' for the reception and relaying of resonance

---

[24.] The small organelles in plant cells that photosynthesize, chloroplasts, were similarly a symbiotic association of a bacterium with a plant cell, much like mitochondria were in animal cells.

information from the morphogenetic background field.[25] Human ATP-driven energy production is consequently not a "heat power machine" but a light frequency modulated information transforming medium. Foods are part of that source of information; how they are grown, what color they are and their rich and diverse interaction with the soil and sunlight. It is a huge shock to realize that the information is not derived just from molecules, that is, matter-driven. This is because light is not a particle but a wave of potentiality (though it can behave as a particle or "photon" when observed). This is quantum mechanical view of the universe and has been known in physics for over one hundred years. Now, if you are scientifically trained like I am, you might wonder why you have never heard of this quantum theory of energy formation.[26] Strangely, biological sciences are still taught via the mechanistic view of chemical reactions occurring with solid atoms and molecules bouncing around into each other. This is out of date, and it severely limits our understanding of health and disease. The answer to why it is not taught may have more to do with politics than with science. It is an "inconvenient truth" maybe. So, choosing food of the best possible quality is not just good for you, it maximizes your health in the long-term.

Herbal Tinctures and Medicinal Herbs

Certain herbs have strong medicinal capabilities, beyond their use in foods. Some, like ashwagandha and ginseng, are *adaptogens*. They are able help your body to adapt to stress by signaling to your adrenals to tune your system *up or down* depending on what is needed. No pharmaceutical has so far been produced that can do this. Drugs are blunt instruments designed to work on limited aspects of your physiology and tend to work on specific pathways in predictable ways. Herbs, having been around a long time, are perhaps more adaptable and intelligent. The level of communication with our microbiome and our gut cells is more complex because herbal extracts contain a multitude of synergistic chemical components—only some of which we have so far identified.[27] By taking "bitters" (an herbal tincture), for instance, more hydrochloric acid (HCl) is produced in your stomach to destroy pathogenic bacteria and increase protein digestive ability.

[25.] This is the vibrational field which is now understood to be everywhere and out of which matter emerges.
[26.] Which hitherto has always been explained as a mechanistic chemical reaction of high energy phosphate bonds being released to drive ATP synthesis as a sort of heat-exchange pump.
[27.] For a good summary of foods and their healing properties, see *The Anti-Cancer Diet Book* by David Servan-Schrieber).

Rebuild the Gut—Basic Protocol

- HCl (from papain) is more than just about improving digestion—it is antimicrobial).

- garlic (not just antimicrobial) stabilizes microvilli, which are hair-like undulations in the gut lining

- CoQ10 (for energy)

- aloe vera (soothing)

- bile salts (help to break up fat properly)

- peppermint (a digestive)

- lion's mane (a very powerful medicinal mushroom [Burke19])

- colostrum and whey powder

- slippery elm powder

- lactoferrin

Be aware that it usually takes at least eight or nine months to heal the gut. And this basic list does not constitute a proper protocol (many of which I describe in detail later). These are products that have helped a large number of people but may not be needed by everyone. The basic idea is always to improve the terrain; first, weeding out the parasites, then reseeding before we then feed, that is, we add probiotics and small amounts of fiber (a form of carbohydrate) to avoid bloating. Fiber often causes a problem with bloating and certain medical interventions like the FODMAPS diet[28] have addressed this by cutting it out. However, this is a dangerous move when it replaces those foods with highly processed alternatives like white bread and rice (which have very little nutrients). It also ignores the important contribution of the brain to bloating because it's partly neurological.

So, in summary, it is not about avoiding foods for life, but about gently rebuilding your gut "under the radar." In this way, we maintain motility of the gut and turn the system back on. Once the gut is healed certain foods no longer stimulate the immune system into self-defense and we are able to tolerate a wider range.

[28.]FODMAP: An acronym for fermentable oligosaccharides, disaccharides, monosaccharides, and polyols (forms of short-chain carbohydrate that are poorly digested in the gut when your gut bacteria are out of balance).

Turmeric

Curcumin is the active ingredient of turmeric—that bright orange spice used in Indian cooking. It is, however, much, much more than a culinary herb. It is one of the most potent anticarcinogens (anticancer molecules) in nature and there is much research into its medicinal use, either on its own or a purified version to be developed into a drug, particularly those encapsulated in tiny particles (so-called "nano-medicine") [Bilia17], [Quispe22]. Interestingly, its vibrational quality is similar to one of the protein complexes in the mitochondrial ETC, Cytochrome C; it emits radiation at a similar frequency and may help to make up for the lack of capacity within that energy producing complex as we age.

Turmeric has also been called "the master conductor of inflammatory pathways" by Jeffrey Bland, a world-leading expert on autoimmune disease. It has the capacity to "tune" our nervous system, so that it becomes healthier; specifically it prevents the destruction of the myelin sheath around nerve fibers "by inhibiting neuronal apoptosis (programmed cell death)" [Yu16, Abstract, p. 1]. "Due to its positive influence on brain-derived neurotrophic factors (BDNF—important for nerve cell growth) and metabolism of amyloid plaques" [Kalinik16, para. 4], it is powerfully anticancer, anti-inflammatory and seems to be beneficial in a wide range of diseases from Alzheimer's and diabetes through to arthritis and heart disease. It is a true superfood. But a bit of advice; it works best when partnered with pepper for absorption, so either cook it with pepper and oil (it's fat soluble) or, if supplementing with tablets, make sure the supplement has piperine (the active ingredient in pepper) included.

There are many other herbs and their active components which have potent effects on the microbiome:

- cloves—contain eugenol which helps to "strengthen the mucosal barrier by increasing the thickness of the inner mucus layer, which protects against invading pathogens and disease" [Wlodarska15, para. 1]

- cinnamon – cinammonaldehyde

- oregano – carvacrol

- fennel – anathol

Nutrition for Brain Health

With the understanding of the gut and the brain as one continuous system, we see that they are intimately connected, and we cannot "fix" one without the other. We have seen how the *brain- gut-microbiota (BGM) axis* communicates between all three organs to regulate immunity and metabolic function. The brain is a key component of this axis and has a high nutrient requirement. Therefore, any problem with energy production will upset the brain, muscles, and heart disproportionately, probably in that order.

The role of mitochondrial dysfunction, which we have described already, has a huge role in memory and cognition. So cognitive decline doesn't exist in isolation, there will always be digestive issues, muscle aching, and brain fog too. A recent paper identified mitochondrial dysfunction is involved in dementia and other neurodegenerative diseases which up to now has always been considered a brain issue [Wang20]. Despite millions of research dollars of investigation this approach has not yielded any significant breakthrough treatments for this condition as a result.

Supplements for Brain Health

One of the major supplemental interventions we can make is increasing our vitamin D3. The optimal range is 80ng/ml. or 2,000–4,000 IU's. Our hunter-gatherer forebears were in the sun without clothing day after day, making vitamin D from sunlight in their skin; our modern lifestyles bear no relation to this. We often get no exposure to sunlight at all if we work in offices and drive in cars. Vitamin D does not get created in the skin when sunlight goes through glass.

In point of fact, it has been misnamed; it is not a vitamin at all, as the definition of vitamin means something that the body cannot make and has to be ingested (which we have seen is not the case unless you have no sunlight exposure). Vitamin D needs a reassessment—it is actually a *prehormone* that is involved in a vast array of processes including immunity, cardiac, and bone health. Lack of it has generally been shown to be "associated with an increased risk of type 1 diabetes (T1D), cardiovascular disease, certain cancers, cognitive decline, depression, pregnancy complications, autoimmunity, allergy and even frailty" [Hossein-nezhad13, p. 720]. The mechanism has not been fully determined but seems to be linked to the interaction of the vitamin D receptor (VDR) in immune response and gut homeostasis. According to a recent study, "evidence shows that hormonal compounds and by-products of the microbiota, such as secondary bile acids, might also activate VDR" [Clark16, para. 1]. Supplementation has been shown to very effective in the treatment of MS and uterine fibroids. It is an easy change to make to your diet as you can find fish oil or algal (vegan) sources in capsules or drops.

You can combine this with certain good unsaturated fats like alpha-lipoic acid (ALA) and DHA/EPA (1000mg/day—algae or fish derived), both of which are used heavily by brain cells in their membranes and often lacking in modern diets. Vitamin B12 is also a good vitamin for the brain, often deficient and deserves to be supplemented, particularly if you are vegan.

References

[Acin-Perez10] Acin-Perez, R., Hoyos, B., Zhao, F., Vinogradov, V., Fischman, D. A., Harris, R. A., Leitges, M., Wongsiriroj, N., Blaner, W. S., Manfredi, G., and Hammerling, U. "Control of oxidative phosphorylation by vitamin A illuminates a fundamental role in mitochondrial energy homoeostasis," *The FASEB Journal* (2010): *24*(2), pp. 627–636.

[Bilia17] Bilia, A. R., Piazzini, V., Guccione, C., Risaliti, L., Asprea, M., Capecchi, G., and Bergonzi, M. C. "Improving on nature: The role of nanomedicine in the development of clinical natural drugs," *Planta Medica* (2017): 83(5), pp. 366–381.

[Brisson20] Brisson, J. "Is inulin good for dysbiosis? A cautious recommendation." Fix Your Gut, July 23, 2020. https://www.fixyourgut.com/cautiously-recommend-fos-inulin/

[Burke19] Burke, V. "Lion's mane mushroom—Unparalleled benefits for your brain and nervous system." GreenMed*info*, July 31, 2019. http://www.greenmedinfo.com/blog/lion-s-mane-mushroom-unparalleled-benefits-your-brain-and-nervous-system

[Clark16] Clark, A., and Mach, N. "Role of vitamin D in the hygiene hypothesis: The interplay between vitamin D, vitamin D receptors, gut microbiota, and immune response," *Frontiers in Immunology* (2016): 7, p. 627.

[Enig00] Enig, M. G. *Know Your Fats.* Bethesda Press, 2000.

[Erickson17] Erickson N., Boscheri, A., Linke, B., and Huebner, J. "Systematic review: Isocaloric ketogenic dietary regimes for cancer patients," *Journal of Medical Oncology* (2017): 34(5), pp. 72.

[Gates11] Gates, D. The Body Ecology Diet. Hay House, 2011.

[Gkogkolou12] Gkogkolou, P., and Böhm, M. "Advanced glycation end products: Key players in skin aging?" *Dermato-Endocrinology* (2012): 4(3), pp. 259–270.

[Hertz15] Hertz L., Chen, Y, and Waggepetersen, H. S. "Effects of ketone bodies in Alzheimer's disease in relation to neural hypometabolism, β-amyloid toxicity, and astrocyte function," *Journal of Neurochemistry*, (2015): *134*(1), pp.7–20.

[HHP20] "Butter vs Margarine." Harvard Health Publishing, January 29, 2020.

https://www.health.harvard.edu/staying-healthy/butter-vs-margarine

[Hossein-nezhad13] Hossein-nezhad, A. and Holick, M. F. "Vitamin D for health: A global perspective," *Mayo Clinic Proceedings* (2013): 88(7), pp. 720–755.

[Juárez15] Juárez-Hernández, Chávez-Tapia, N. C., Uribe, M., and Barberro-Becerra, V. J. "Role of bioactive fatty acids in non-alcoholic fatty liver disease," *Nutrition Journal* (2015): *15*, article 72.

[Kalinik16] Kalinik, Eve, "Nutrition Notes: Tune in to turmeric," Available online at:

https://www.psychologies.co.uk/nutrition-notes-tune-turmeric

[Kopec10] Kopec, R. E., Cooperstone, J. L., Schweiggert, R. M., Young, G. S., Harrison, G. H., Francis, D. M., Clinton, S. K., and Schwartz, S. J. "Avocado consumption enhances human postprandial provitamin A absorption and conversion from a novel high– -carotene tomato sauce and from carrots," *The Journal of Nutrition* (2014): *144*(8), pp. 1158–1166.

[Lakhan10] Lakhan S. E. "Gut inflammation in chronic fatigue syndrome," *Nutrition & Metabolism* (2010): 7, article 79.

[Lewis17] Lewis, T. im, "Meet the chef who's debunking detox, diets, and wellness." *The Guardian*, June 18, 2017.

https://www.theguardian.com/lifeandstyle/2017/jun/18/angry-chef-debunking-detox-diets-wellness-nutrition-alternative-facts

[Longo18] Longo, V. *The Longevity Diet*. Penguin, 2018

[Longo22] Longo, V. "Periodic fasting mimicking diet, longevity, and disease," *Innovation in Aging* (December 20, 2022): 6[Suppl. 1], p. 91.

[Martin16] Martin K., Jackson, C. F., Levy, R. G., and Cooper, P. N. "Ketogenic diet and other dietary treatments for epilepsy." *Cochrane Database System Reviews* (February 9, 2016).

[Nordmann06] Nordmann A. J., Nordmann, A., Briel, M., Keller U., Yancy Jr., W. S., Brehm, B. J., and Bucher, H. C. "Effects of low-carbohydrate vs low-fat diets on weight loss and cardiovascular risk factors: a meta-analysis of randomized controlled trials." *Archives of Internal Medicine* (2006): *166*(3), pp. 285–293.

[Pace15] Pace F., Pace, M., and Quartarone, G. "Probiotics in digestive diseases: Focus on Lactobacillus GG," *Minerva Gastroenterologica e Dietologica* (2015): *61*(4), pp. 273–292.

[Paoli13] Paoli, A., Rubini, A., Volek, J. S., and Grimaldi, K. A. "Beyond weight loss: A review of the therapeutic uses of very-low-carbohydrate (ketogenic) diets," *Journal of Clinical Nutrition* (2013): *67*(8), pp. 789–796.

[Prado15] Prado, M. R., Blandón, L. M., Vandenberghe, L., Rodrigues, C., Castro, G. R., Thomaz-Sokol, V., and Sokol, V. R. "Milk kefir: Composition, microbial cultures, biological activities, and related products," *Frontiers in Microbiology* (2015), *6*. https://www.frontiersin.org/articles/10.3389/fmicb.2015.01177/full

[Quispe22] Quispe, C., Herrera-Bravo, J., Khan, K., Javed, Z., Semwal, P., Painuli, S., Kamiwoglu S., Martorell, M., Calina, D., Sharifi-Rad, J. "Therapeutic applications of curcumin nanomedicine formulations in cystic fibrosis," *Progress in Biomaterials* (2022): *11*(4), pp. 321–329.

[Singh14] Singh, V. P., Bali, A., Singh, N., and Jaggi, A. S. "Advanced glycation end products and diabetic complications." *The Korean Journal of Physiology & Pharmacology* (2014): *18*(1), pp. 1–14.

[Velasquez15] Velasquez-Manoff. M. "Gut microbiome: the peacekeepers," *Nature* (2015): *518*(7540): pp. S3–11.

[Wang20] Wang, W., Zhao, F., Ma, X., Perry, G., and Zhu, X. "Mitochondria dysfunction in the pathogenesis of Alzheimer's disease: recent advances," *Molecular Neurodegeneration* (2020): *15*(1), article 30.

[Wlodarska15] Wlodarska M., Willing, B. P., Bravo, D. M., and Finlay, B. B. "Phytonutrient diet supplementation promotes beneficial Clostridia species and intestinal mucus secretion resulting in protection against enteric infection," *Scientific Reports* (2015): *19*(5), article 9253.

[Xu14] Xu, C., Zhang, J., Mihai, D. M., and Washington, I. "Light-harvesting chlorophyll pigments enable mammalian mitochondria to capture photonic energy and produce ATP," *Journal of Cell Science* (2014): *127*(2), pp. 388–399.

[Yu16] Yu, H. J., Ma, L., Jiang, J., and Sun, S. Q. "Protective effect of curcumin on neural myelin sheaths by attenuating interactions between the endoplasmic reticulum and mitochondria after compressed spinal cord injury, " *Journal of Spine* (2016): 5(4), pp. 1–6.

CHAPTER 5

BEYOND FOOD: LIFESTYLE FACTORS

Lifestyle Changes

There are many lifestyle factors that ameliorate threats in your gut flora through epigenetic effects. Modifiable lifestyle factors are one way we can overcome the threat of disease—even the most feared of modern diseases cancer and, especially, the autoimmune type diseases—both of which we are losing the battle with currently. These chronic inflammatory diseases are testing our health services in unprecedented ways that we simply don't have treatments for.

There has been a growing recognition that lifestyle might have something to do with this rise in *noncommunicable disease* (NCD). A return to a primal lifestyle has been mooted as a solution; also known as the "Paleo" movement. It has been billed as being primarily about changing what you eat but, in fact, is not all about diet. It takes in exercise/outdoor engagement, relationships and connection, managing stress and so on. We need to get back to lifestyles that are more conducive to health and off the tightrope of chronic disengagement from our place in the ecosystem of the planet and our purpose as human beings. This is what primitive man had instinctively.

Food, exercise, and stress are some of the most powerful epigenetic modifiers. We will look at each in turn, as well as some not so obvious ones. Food being the most powerful of the three, has a chapter all to itself (Chapter 6), but we start now with the others.

Movement/Exercise

We are designed to move. It has long been noted that athletes have very low incidence of gum disease (a precursor to gut dysbiosis). Could it be their high exercise rates that reduces their inflammation? It seems this is the case—in fact bouts of exercise seem to encourage neurogenesis (creation of new neurons and neuronal connections) [Hunt12].

But be careful too not to overdo it—particularly if you have energy issues. It's a fact that over-exercising (especially compulsively) can, for women particularly, accelerate aging and cause hormonal difficulties that stop their periods (amenorrhea). According to recent research, this is because "unaccustomed and/or exhaustive exercise can generate excessive reactive oxygen species (ROS), leading to oxidative stress-related tissue damages and impaired muscle contractility" [He16]. If you don't work up to it gradually and make sure your diet includes increased amounts of antioxidants (vitamins C, E, glutathione, and CoQ10, among others) [Belviranli15], your body cannot "mop up" the free radicals and ROSs produced in the mitochondria when we exercise. There is also the psychological stress factor to take into account; if you are overdoing it because you fear getting fat or believe you are a failure if you don't exercise, your body interprets this as a stress and excess cortisol is produced.[1] This, as we have already indicated, is not good for your long-term health. The question to ask yourself is: if I don't exercise do I feel anxious? If so, you may have an addiction/compulsion and this is something you need to address.

Get into the flow! Exercise is known to have important mood-altering properties and to be highly beneficial in depression, for instance. But how does it do this and what type of exercise is best? Well, there are many theories as to whether aerobic or strength exercise is best; I tend to think a mixture is ideal, with more of the resistance exercise as you age and muscle mass tends to decrease. I like yoga or Pilates for this reason, but some people don't care for these, so just choose the one that you will do. In other words, pick something that you enjoy, so it doesn't feel like a chore that will invariably be dropped when other more important tasks come up. One with a social element is particularly important if you are depressed. So, for example, join a running or walking club (some GP surgeries now have these and "prescribe" walking instead of a pill—a commendable decision). Martial arts/yoga are good for people who have low self-esteem as they

---

[1.] See https://www.psychologytoday.com/blog/fit-femininity/201504/can-we-exercise-too-much for more info.

build body sensation (*interoception*) and develop a sense of internal safety (*neuroception*).

It is more beneficial to do your exercise outdoors in sunshine—for vitamin D and natural light, but whatever you find that works for you is better than the exercise that you won't keep up. I used to go to the gym a lot when I was in my 20s and 30s, but I found I was "rewarding" myself with high carb meals afterwards, so even though I developed good muscle tone, I'm not sure overall that I was benefiting.

Moderate exercise (at least three times a week) is good for the heart and cardiovascular system helping to promote healthy blood flow and mitochondria to renew. It has even been shown to improve mood and to be more beneficial in mild depression than antidepressants [Blumenthal12], especially if combined with mindfulness training [Alderman16]. Exercise is a powerful *epigenetic modifier*, that is, it can change your gene expression. This is particularly true if you exercise each day (approximately twenty minutes of aerobic exercise per day). In order to be beneficial, you need to get your heart rate up to 80% of your maximum, for example, 180 bpm—dependent on age and medications (beware beta blockers reduce heart rate). I have found a short jog in the morning three times per week that takes me less than twenty minutes, works for me—if combined with cycling to work on the days I don't run. This is because I don't have to make myself; I enjoy it and save money in the process. You will have a different experience but be aware it is one of the most important and underrated interventions you can make in your life. And it doesn't have to be onerous— try burst (interval) training where you do three rounds of three to four minute aerobic exercises with periods of "cool down" in between are most effective—much more so than an hour on a treadmill. This is because it boosts human growth hormone (HGH),[2] which despite its name isn't just for our childhood years, but it's also an important regulatory "feel good" hormone throughout life for both men and women, but particularly in adolescence [Cockcroft15].

The importance of exercise for our discussion is that it *inhibits inflammatory processes* of cognitive decline, joint swelling, cardiovascular clogging, and so on. How it does this is not altogether determined yet, but it seems to "upregulate" certain protective molecules and promote the detoxification process within the liver–both of these involve *methylation*

[2.] HGH is produced in the pituitary and stimulates growth, cellular renewal, and regeneration; and boosts fat-burning via its adrenaline (epinephrine) release.

which is the addition of a methyl group (CH3–carbon and 3 hydrogens) to DNA or other molecules. As discussed in more detail in Chapter 3, DNA methylation changes the readout of a DNA segment without changing the sequence (i.e., genetic code). It is thus a way of modifying our biochemistry much more quickly than genetics would be able to do (as they rely on a longer process of reproduction). Hence, it's like having a "fine-tuning" ability rather than just an on/off switch.

Sunlight Exposure

Adequate sun exposure is vital. The most important thing that we gain from sunlight is vitamin D which is made in the skin from sunlight of a certain frequency. Most people in the western world lack enough vitamin D due to our indoor lifestyles; deficiency is endemic. We cannot make enough even if we do go outside as, during the winter months in the northern hemisphere, the wavelengths are not of the correct length—in the UK we only make vitamin D reliably between April and October. As a powerful prehormone involved in myriad reactions in the body, getting your vitamin D status up to optimum is vital.

It is not only vitamin D that sunshine provides. Infra-red rays restructure water in your body to have a more harmonious electro-chemical structure that mimics the energy of our own body (which is why infrared saunas are so good for detoxing). I will talk about this later. They also make us feel good. (Summer holidays have this effect not only because of relaxation). Light helps balance the pituitary too which is the master hormone gland. So, exercising in the sun is better than in the gym. If you are into running, make it a practice to get out in the air not on the treadmill.

Pure Water Availability

In the developed world we have some of the cleanest drinking water on the planet. But this comes at a price. Clean does not necessarily mean good. Certainly, it means free of pathogenic bacteria that could give us food poisoning or typhus.[3] But in order to render previously used" water (and this can mean used for sewerage, industrial cooling, and of course

---

3. Dr. John Snow solved the cholera epidemic of 1854 by tracking the source of the infection to one water pump in London and removing the handle. He is considered the "father of epidemiology" (study of populations).

drinking), we have to put many chemicals into the water to render it potable (drinkable). It is strange that we still do not separate out drinking water supply from that used to flush toilets for instance (although some people recycle "grey" water for this purpose). It may surprise you that water that comes out of the tap will have upwards of 200 volatile organic chemicals (VOC's) added to make it clean including trihalomethanes (THMs), bisphosphonates, and so on. It will also have a number of natural contaminants (metal salts and nitrates) making it potentially quite hazardous if not treated properly [Fawell14].

When we were living natural Paleolithic lives we would have obtained our drinking water naturally filtered by rock from natural rainfall (obtained from underground wells or streams). In addition, having moved along a watercourse it would have picked up a lot of positive ions and be electrochemically "living." Tap water bears no relation to this type of water other than the same chemical signature H_2O. Its energetic signature is entirely different.[4]

Therefore, the best solution then is to filter your tap water with a good (carbon) filter. The best are reverse osmosis systems which fit under the sink in the kitchen and filter as you run the tap. They are expensive, however, so I have a charcoal system in jugs that makes the water taste much nicer as charcoal attracts heavy metals onto its surface. However, there might be a problem in that charcoal can filter out the good minerals as well as the bad chemicals. In this case you need to supplement. There seems no easy solution at the moment unless you happen to be lucky enough to live somewhere where there is natural spring water. Residents of some spa towns in Derbyshire and Yorkshire in the UK can access naturally filtered groundwater from certain places, but for the rest of us, especially those living in urban environments, this is not possible, and we are dependent on bottled or tap water. Luckily technology is developing all the time and now there are systems which fit directly onto the faucet or tap like that shown in Figure 5.1.

[4.] If you find this difficult to believe, then you will be amazed to find that we can intervene in the quality of our water simply by intention; the Japanese researcher Masaru Emoto did some experiments with water crystals and human interaction; he found vast differences between those subject to loving versus hateful thoughts! Interestingly this scientist was a skeptic before he did the experiments. His results are best seen in the photographs from his bestselling book *The Hidden Messages in Water*.

Figure 5.1. Direct water filter.

Good Sleep Quality

As we have already intimated, sleep is really important for health, good quality sleep particularly. However, that is getting increasingly difficult to achieve in our hyper-stressed, 24/7 access world. So, one aspect that you need to look at is sleep hygiene; keeping light out of the bedroom—particularly blue light such as that from screens—these upset our day/night cycle by stimulating the brain to think it is daylight and therefore not time to sleep. We have already explained how that upsets your circadian rhythm by interfering with your melatonin production. So, installing blackout curtains can be a life saver; they are used in northern climates in the summer to promote sleep during their light evenings.

One simple thing you can do is take electromagnetic fields (EMFs) out of the bedroom—turn off your router—or any wireless connectivity device when you are asleep—or add biofield protection to yourself and your modem. I have a Biodot™ on my laptop, modem, and smartphone, and I wear a Bioband™ on myself when out and about. These EMFs are particularly damaging when the brain is asleep as the electromagnetic wave pattern of the brain alters in sleep and it seems more closely matches that of electronic devices than when awake. So, if you do nothing else, make sure you keep your bedroom clean of EMFs.

Grounding

Another simple intervention you can make to your lifestyle is grounding (earthing): rubbing your hands in soil or walking barefoot, for instance. Be in touch with the earth with socks off occasionally—if the weather allows. The surface of the earth has a slight negative electric charge due to its free electron field. Believe it or not grounding—especially walking on the ground with direct contact of your skin—allows you to use these free electrons to balance the positive charge you carry. It is an increasing problem that we don't discharge.

According to the "father of grounding," Dr. Clint Ober, "unfortunately, with our modern rubber or plastic soled shoes and insulated sleeping arrangements, we no longer have a natural electrical connection to the Earth, unless walking barefoot."[5] Natural rubber was more conductive as it allowed moisture from our sweat to permeate and therefore the water conducts. Modern plastic-derived rubber is not permeable in the same way. Thus, we are largely insulated from the earth and now surrounded by positive charge from pollution, including electromagnetic smog, trans fats in our food, and the free radicals produced from metabolic processes (particularly via the immune system that uses them to kill pathogens). These are all stresses on the body. Is it any surprise we are seeing a huge rise in disease as a result?

So how does grounding work? Authors of a recent paper have suggested a potential mechanism could be "mobile electrons create an antioxidant microenvironment (...), slowing or preventing reactive oxygen species (ROS[6]) (...) from causing "collateral damage" to healthy tissue. We also hypothesize that electrons from the Earth can prevent or resolve so-called "silent" or "smoldering" inflammation" [Oschman15] (what I refer to as systemic inflammation; the heart of most chronic illness). Besides reducing inflammation, it has also been demonstrated to thin your blood, and improve your immune response. It also has this become an issue? It also seems to help the nervous system and shift it to a parasympathetic state; so called "vagal stimulation." It also changes the viscosity of blood making the blood cells free from clumping; this is huge factor in reducing strokes and heart disease. See the videos by cardiologist Dr. Sinatra and York Cardiologist for more information.[7]

The effects are so beneficial and vast [Chevalier12] that you wonder why you haven't heard it from your GP or that the health system doesn't

---

[5.] See http://www.groundology.co.uk/about-grounding for more information.

[6.] ROS are a normal part of the inflammatory response. but when unbalanced by antioxidants are like a wildfire spreading throughout the body unchecked. It seems free electrons provide the "water" to put out the "fire."

[7.] See https://www.drsinatra.com/ and York Cardiologist on YouTube.

subsidize grounding pads for people with chronic illness. You can buy grounding devices to connect to your bed so that you ground while you sleep, or you can purchase small pads to place under your feet or computer while you work; they plug into a power outlet that fortunately is connected to the earth pin in the electrical plug. (I'm using one now while I type.) They bring your voltage down to zero by allowing you to ground to the earth's charge. This is particularly important the higher you sleep above the earth; if you live in a tower block, or flats above the ground your voltage is likely to be even higher [Ober14]. Note that whenever you are in connection with an electrical device (like your laptop for instance) your voltage rises immediately. You can test this for yourself with a home voltmeter kit.

You can obtain under bed sheets too. This reduces your voltage to negligible levels which is so important if you work or sleep with electrical equipment around you. If you suffer from hormonal/sleep difficulties, this could be something to consider as a simple, fairly low-cost method of improving your health. You could say we have an "electron deficiency syndrome." The influence of electric charge on us is something that we would do well to take seriously. The effect of being in contact with positive charge day in day out cannot be healthy. And it is increasingly the case that most people are disconnected from the earth,[8] which we were never designed to be.

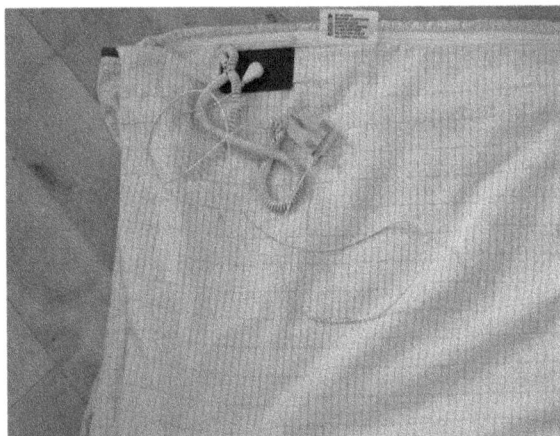

Figure 5.2. Under bed grounding sheet.

---

[8] For the full information watch the video with Dr. Clint Ober and Dr. Mercola on the Groundology website: https://www.groundology.com/ie/

Detoxification and Rebuilding the Gut

We are surrounded by chemicals in the air, water, and foods that we eat, most of them untested on humans and of dubious safety. Our natural detoxification processes (mostly in the liver) are likely to be overwhelmed. What can we do?

You may have been on detox regimes to help clear the body. These are good, but they only last as long as your regime. How do you make changes that last? Well, the good news is that Lactobacillus species helps both to break down *and* to help the body break down heavy metals in the diet and eliminate them in the stool. So, adding these to your gut flora is a good start in detoxification and if you create the right conditions for them to proliferate you will have a longer-term solution to the problem of heavy metals. Unfortunately, if you have ever had a dose of broad-spectrum antibiotics (as most of us have) you are likely to have lost the diversity of your gut flora, and lactobacillus being one of the most prolific in the large intestine, is likely to have been denuded.

Detoxification Protocol

According to "WellnessWiz" Dr. Jack Tips, there is an order in which to tackle detoxification which has the most benefit:

- Build digestion.

- Clean the colon.

- Support the liver.

Many people make the mistake of running straight into detoxification without supporting the body systems first. The inevitable result is—you feel lousy. I myself have done this and had three days of headaches, shivering and nausea when doing a juice cleanse. These days I do it much slower. It is good to do a liver/colon cleanse twice a year—once after Christmas and once after summer vacation. This helps to keep your digestive tract in good condition. After this, you can look at cleansing the extracellular matrix via kidney support and clearing out the lymphatics and neuro-endocrine systems with pituitary support. This then passes on to other organs and glands like the pineal, thyroid, sex organs, and so on. It's really like doing a full service on your car. But if you do nothing, don't be surprised when you clog up and get an illness as a result. Disease is a systemic breakdown, a sign of poor maintenance not a random failure of individual organs or glands. This is really the difference in outlook between natural and conventional views of health.

Supporting the Mitochondria

If you have a chronic illness, you might need to also support your cell mitochondria to allow clearance of toxins from the extra-cellular membranes via the liver and blood. You can't rush the process, and you must respect your body's integrity and natural way of doing things. Those who rush into detoxification may initially feel good when they clear out, but the issue will come back when they don't support the body properly. Also, if they have adrenal fatigue issues (as chronic stress tends to encourage), detoxification itself will be identified by the body as a threat and you will feel worse [Golan98]. Don't forget, to reduce inflammation in the cell membrane you will need antioxidants and chelators to neutralize those toxins (both internal metabolic waste and environmental toxins like mercury).

So, you really need to consider a protocol to support your mitochondria such as this:

- Reduce inflammation (removing parasites and other sources).

- Repair cell membranes (omega-3 fats).

- Boost ATP energy production.

- Provide vitamin B/folate methyl donors—transmutes (neutralizes) toxins and repairs DNA.

- Repair free radical damage. Provide antioxidants, for example, Selenium for super-oxide dismutase (SOD—one of the free radical repairing enzymes) and glutathione as supplements.

- Rebuild gut the flora.

You have to work to this level of clearance if you want to live a long and healthy life. As Dr. Tips says, "We live and die at the cellular level" and this is a profound truth.

Herbs for Parasites and Detoxification

One great way of dealing with parasites and general gentle detoxification is taking certain herbs and bitters on a regular basis. This used to be routine in most people's diets and has sadly declined.

The main herbs proven to have antiparasite activity are:

- berberine

- grapefruit seed extract

- dandelion root

- garlic

- milk thistle (silymarin)

- black walnut

- centaury (Native Americans used this)

- wormwood (artemesia)—contains artemesinin[9]

- olive leaf extract

- turmeric

- basil (especially holy basil)

However, it is worth reiterating that if you have a healthy microbiome the most common parasites can't exist, so prevention is an important part of the program. Foods that contain fermented agents high in lactic acid, for example yogurt, pickles, and so on, are recommended for this purpose. You may add in proteolytic (protein breakdown) enzymes (especially those ones made from fungus) or pancreatic enzymes to break down parasite eggs and a daily dose of vitamin A and zinc.

Antiparasite Protocol

Dr. Hazel Parcells, grande dame of alternative medicine (who lived to 106 with all her faculties intact), was very keen on promoting good food and the importance of cleansing parasites. According to Ann Louise Gittelman, who worked with her for twenty years, the program she offered to draw parasites out was an eight-day milk diet; you drank milk (goats or raw cow's milk and specialty herbs). Milk acted as "bait" because parasites live within the gut wall, and thus you can't see them in a conventional stool sample. The herbs then dealt with the adults and then finally it was adjusted to kill the eggs and larva. There are other methods that work more slowly, and some ready-made parasite cleansing supplements are available. But foods and herbs are the best way of dealing with them. Olive leaf extract is particularly powerful.

Certain foods like cranberry are rich in organic acids that help to break down the parasite protein structure and release the waste. It is suggested to take eight ounces of unsweetened cranberry juice with a large glass of

---

[9.] Warning: this is strong—you must not take on a daily basis due to toxins released; the liver may not cope. Best to do two weeks on and then five days off.

water throughout the day. Native Americans used it as a deworming agent once a week. Also, you can eat toasted pumpkin seeds every day (very high in zinc). Parasites don't do well in high zinc or vitamin A environments. Also take garlic with your foods; the alicin in garlic kills parasites in the intestinal tract—raw or aged garlic are both suitable, cooked is not so good. You could also try cayenne pepper and mugwort tea on a daily basis. They all love instant sugar. So, limit these and all gluten-rich foods and most dairy which are similarly likely to encourage them. A natural fermented vegetable product called kimchi (see Korean food stores), helps to degrade organophosphates to produce energy so it adds to our arsenal of detoxification mechanisms.

Parasites have a massive effect on the microbiome. Imbalance in microbiome and parasites always go together. You should beware of taking any beneficial bacteria before ridding the gut of parasites. This is because, according to Ann Louise, probiotics strengthen certain worms and flukes too; it is best, therefore, to treat parasites for two weeks, then add probiotics.

One way to start is with a colon cleanse (either colonic irrigation or a home cleansing kit). Use the gentle herbs like berberine, grapefruit seed extract, and dandelion root first. This is because if your liver is not performing well, that is, not producing enough bile, it will be unable to digest the protein fragments of the parasites as they die off. You must make sure liver and bile production are well supported whatever protocol you use. If your parasite infection is particularly severe (and you will find out when you do a cleanse), you may need a specific antibiotic for specific parasites in short stints for perhaps five days. Then follow with something more natural.[10]

There are various downloadable online protocols for the natural healing of parasites. Testing to identify what parasites are present may be necessary first. Some people suggest that you can identify by kinesiology (muscle testing) which is another route you may wish to investigate rather than formal stool testing.

Essential Oils

Essential oils are natural volatile organic compounds (VOCs) which help to ward off parasites and heal us from infection/disease.[11] These VOCs

---

[10] For regular cleansing remember parasites are most active five days before and after full moon. Do the protocol then.

[11] Technically, they're not oils at all lacking fatty acids.

include ketones, terpenes, esters, and so on. There are three main methods of application:

- Aromatherapy—for example, burning incense.

- Topical—applied directly onto the skin (but you must dilute most, except lavender).

- Internal—dilute, and take as drops into the mouth

You can make your own products by using a few drops in a carrier oil such as coconut oil, olive oil, or shea butter. The wonderful thing about essential oils is that they are *selectively* antimicrobial—they leave the healthy bacteria and only target pathogenic ones.

Here are some examples of good blends for different purposes:

- Air cleansing: frankincense—kills airborne pathogens; called dhoop in India (used for ritual spiritual purifications).

- Athletes: cardamon, eucalyptus, lemon, peppermint, rosemary, tea-tree.

- Cancer: frankincense—good for skin cancer and all other cancers. Get frankincense resin and burn it every day. You can also apply it topically to skin cancers.

- Focus: frankincense, sandalwood, cedarwood, vetiver. These last two particularly help children focus.

- Immune boosting: cinnamon, clove, eucalyptus rosemary, orange, lemon.

- Joyful: citrus oil like orange, lemon, grapefruit, and lime + vanilla or even bergamot

- Skin: frankincense. Good for skin tags and tumors, warts, and so on. Can also be used in lotion and soap. Clove, thyme, and jojoba oil on blemishes. For eczema: wash with coconut oil and water, then cider vinegar, then 50/50 coconut oil/aloe gel. For chapped lips—coconut oil with berberine (a powerful antimicrobial) as a lip balm and ideally a probiotic added would be good—not available commercially so you will need to make your own.

- Sleep: chamomile, lavender, vetiver.

Essential oils help promote intestinal absorbability of probiotics so aid in general rebalancing. You can add to an unscented soap base with some essential oils to make your own soap too. They really are invaluable—check out some of the specialist websites drericz.com and dr.axe.com.[12]

Relationships and Connection

This seems perhaps an unlikely section in a book about the microbiome, except that maybe it's worth reiterating that our relationship with our microbes is something that we develop more intimately when we become aware of our own place in the ecosystem. In order to truly enjoy our life and thus nurture ourselves we need connection with others of our species (and of course other species too.). As many studies have shown, how we live in relationship to each other has huge effects on our stress levels and thus our health.

Many of us are hardwired as children to be permanently stressed—we may have had emotional trauma or chronic stress in our families of origin; this lowers our stress "set-point" and makes us more likely to have further stress. Our "resilience" is said to be lowered. Since you can't choose your families of origin, then choose easy-going people to hang around with. Cut out the "energy-vampires"[13] and make sure you set up good boundaries with the people in your life. Learn to say "no" occasionally (great for all your people-pleasers out there).

There are some simple things you can do:

- Get involved in acts of appreciation and giving as a part of your daily life. I make a point of mentally thanking the good things in my life when I wake each morning. This may sound rather trite, but it has the effect of changing your brain toward positivity rather than looking for problems, which is how most people are most of the time.

- Cultivating kindness and compassion; this is harder especially with those who irritate and hurt us. But you can start with yourself and your self-talk. Watch and note how you talk to yourself; it's likely very negative and derogatory. Remember this does not build our self-worth[14] and it is with ourselves that love begins. Once you learn to be kinder to yourself, it becomes easier with others. People who hit out at others often do so because it matches what they feel about themselves.

---

[12] Both are good in different ways—the former is more biblical due to Dr. Z's Christian beliefs.

[13] A term used to describe people who drain you by their mood or behavior.

[14] Lack of self-worth will override any dietary changes you make and will eventually mean you abandon them. Looking at your emotional health is supremely important. See my previous book *The Scar that Won't Heal*.

▪ Work on your mindset as it has everything to do with the trajectory of your life (both morbidity (illness) and mortality). If you have a victim mentality it becomes a mental habit to always see things in this light, and your life will then reflect this back to you. Become "miracle-minded" and cultivate positivity; mindfulness has a great part to play in this.

▪ The more you activate drive and will toward finding meaning and the needs of others, that is, something that transcends your own experience, the more focused you become on connection which is one of the most profound indicators of longevity. Volunteer, get active.

▪ Avoid negativity—whether from news, people, organizations or other sources, and actively choose positive people, situations, and jobs when possible. Sometimes this will be about changing your job or friendship network; other times it could be as simple as signing up to a positive news channel.

▪ Change your world view toward resolution not problems, that is, a "things will get better" mentality. The microbiome picks up this message. How you do this is covered in more detail in my first book, *The Scar that Won't Heal* but it often involves clearing old belief systems.

In short, you are looking to create cohesiveness, balance, and unity in your life, either through a spiritual practice of some kind or a passion for something that you do uniquely well. This is the essence of holism. I know some of this will sound very new-age to some of you, but if I can just say that the rewards are very practical: you will improve your diet, but it will also work on your relationships and every aspect of your life. This is what will keep you and your microbiome healthy in a reciprocal relationship. This is the basis of the Chrysalis Effect Program of which I am a specialist in—we work with CFS/ME specifically, but the approach applies to all chronic illness. Clients are asked to look at eight elements of their lives including diet, lifestyle, environment, and so on as it is necessary to work on all of these in order to recover from such complex conditions. But there is no reason that you could not do this preventatively; it's just that most people will not act before they are forced to by illness.

As a society, we need to eliminate the pessimistic, ageist thinking that everything goes downhill as you age. I can't tell you how many times I have come across people in terrible pain who believe it is their age and nothing can be done—this is not true. In part, this is the result of our cultural bias against age. But, also, it is about fear which is promulgated by most news media and even our medical institutions; for instance, the seemingly inevitable loss of income, meaning, and connection when we get old. But this is not inevitable *in a healthy society.* Yes, the physical aspects of your body may get weaker (although in healthy individuals this is minor). But there are gains too. The spiritual knowing that one gains from aging are seldom talked about. Provided we have managed to build a life of connection,[15] we can reduce our sense of separation from others and actually enjoy old age. Remember it is the separation from the whole that poisons our worldview. This is the teaching of ancient traditions and also, happily, corroborated now by modern science. Restoring our connection is the prime directive if we wish to age well. Get out in the fresh air, make friends, join communities, laugh a lot. And if you are ill and can't get out, do it online—but beware the negative groups who are addicted to their own illnesses.[16] Be very selective in which support groups you join.

Health entrepreneurs, such as JJ Virgin, talk about maintaining a "miracle mindset" when bad things happen [Virgin16]. Believing that things will work out and give that the energy, not the worst-case scenario. It is *not the avoidance of bad things that allows you to develop resilience,* but to believe you can *handle* anything that happens by developing tools and skill sets to replace the habit of worrying/rumination and giving energy to the negative. "Leaning into fear" builds your resilience muscle. Start small and then increase your abilities; this can be greatly helped by working with someone who has gone before you on the same path. You can't heal your body without changing your mindset. Get professional help with this if, like most people, you find it difficult.

Remember, if you feel resistance, know that fear is at the base of it. Resistance shows up in order for you to learn how to deal with it, not to avoid it. It also gives you a clue that something is important to you. Step out of your comfort zone. Reframe fear as a challenge/opportunity to stretch. This expands your comfort zone; a process which is irreversible so you can't go backwards. Remember: "little hinges swing big doors," as the old adage goes.

The current way we live our lives, with its separation from each other, ourselves, and nature is a dangerously false construct. Nowhere is this truer than in modern medicine. Look at the average hospital for instance;

[15.] Many studies show that all the longest-lived cultures have social connection at their heart. There is a direct correlation of heart disease and loneliness, for instance.

[16.] This is not just a euphemism; when we are ill, we can get our significance from being ill as everything else has been stripped away. We then get a dopamine rush when we find corroboration of our illness—this is truly addiction.

it is shocking when you go to visit and you are confronted by junk food, dirty wards, and no art or gardens. At my local university teaching hospital (which is regarded as one of the better ones in the south of England), the outside spaces are crammed full of smokers enjoying an illicit cigarette and polluting the air for everyone else (sorry smokers). There is nowhere to sit and contemplate nature, which would be so restorative. Everyone is either rushing in or out, there are bright fluorescent lights, clashing noise, fast food joints, and it is soulless. I have never seen a worse place to heal. And the rate of infections bears testament to this. Hospitals are often the worst places to be ill. Hospices are often much, much better but, of course, they are the final stop on most people's road. Wouldn't it be great if we put as much effort into healing before we get so ill that it is too late? I leave you to contemplate that.

Life Stages

Now I wish to consider the special features of the different stages of life and how they present specific issues. I start with the beginning of life.

Conception

It may seem strange to you that I start with such an early part of life; after all isn't a human being at conception still so tiny, so nascent, and lacking consciousness that it can't have any bearing on the life of the future person? Not true. According to Dr. Wendy McCarthy, the latest research in prenatal and perinatal Psychology (PPN) shows us:

- We are conscious, aware, learning intensely and actively communicating and forming relationships from the beginning of life (i.e., conception onwards).

- Our earliest experiences in the womb, during birth and bonding, and as young babies profoundly shape and set in motion physical, mental, emotional, and relational life patterns that can be life enhancing or diminishing.

- The way we are conceived, a process called epigenetic imprinting adjusts the activity of genes that shape the character of the child yet to be born. Thus, the state of mind of our parents at conception also alters the life trajectory.[17]

[17.] As Bruce Lipton demonstrated in his book "The Biology of Belief."

So, we have to consider the influence of the microbiome on fertility prior to conception; if a woman's body is stressed and out of balance, it is unlikely that getting pregnant will be easy. Indeed, the former UK-based charity organization Foresight, which promoted natural approaches to pregnancy and especially "preconceptual care," believed that it was important to establish digestive health because the gut is so linked with fertility. This is because "infections of any type can decrease pregnancy chances, so establishing a healthy amount of good bacteria can help support the immune system, as well as supporting the vaginal ecosystem to promote a better chance for sperm to be able to travel." Moreover, it's not just women who are affected, of course: "new research showed men who were put on probiotics to establish beneficial strains of bacteria in the gut showed a higher sperm count, a higher count of sertoli cells which are responsible for testosterone levels, elevated testosterone, and more vigorous ejaculum" [Weng14, Abstract]. It is vital that both partners assess their gut health and make changes if it is lacking; this is particularly important if infertility is already an issue.

Pregnancy

The quality of a pregnancy determines future health of child in many ways; some would say it is the most important determinant for the life of the adult.[18] In order to reduce the risk of miscarriage, the mother's body must adapt its immune response so that it does not reject the foreign tissue of the baby at crucial point. But the latest understanding is that we initially need a proinflammatory response in order to implant in the womb lining, which then switches to an anti-inflammatory one, before finally switching again prior to birth [Aghaeepour17]. Much of this is mediated by our microbiome.

The last trimester is largely a proinflammatory (Th1) state without which the mother cannot go into labor.[19] This is mediated by the microbiome and signaling molecules that they create, which is transmitted to the fetus via the placenta. So, anything that interferes with this signaling, administration of antibiotics, for instance, will more likely have an altered immune response, for example, an atopic (Th1 or "allergic") baby because her natural microflora is killed off indiscriminately. In addition, according to the "double-hit hypothesis," viruses that would normally be kept in check

---

[18.] There is a hypothesis, called the "Barker hypothesis" that the 'developmental origins of health and disease' (DOHAD)that is, what happens to the fetus in utero are primary influences in health and disease.

[19.] And indeed, failures of this switch can force pre-term delivery.

by the bacterial balance can make "the mother at risk of bacterial infection in addition to the virus already in the system" and this can cause pre-term birth [Hewings-Martin17]. This destruction of the natural balance of microbes within her dictates the baby's gut microflora upon birth and thus determines the future health of that child. The microbiome of the mother affects (primes) the immunity of the baby's gut. It used to be thought that the baby's gut was sterile until inoculated by vaginal flora at birth, but new analysis reveals the baby's first stool has traces of the mother's microbes in it so there must have been some transfer.

Babies are selectively nourished compared to their mother, that is, they take all her available nutrients (and, perhaps surprisingly, her toxins too). While pregnant, the mother therefore needs to take care of her diet by taking good probiotics, especially Bifidobacterium; and by eating green foods. It goes without saying she should also take care of her stress levels as this affects the microbiome directly via stress hormones.

Birth

Birth practices have changed in unprecedented ways in the last twenty to thirty years. The march toward medicalized birth began much earlier in the last century, but more recently the rise in elective Caesarean (C-section) births and lack of breast feeding, are beginning to have profound effects on our microbiome. In a "normal" (vaginal) birth, the baby is inoculated with the mother's gut flora as it comes through the birth canal. When a baby is born, the baby swallows and inhales mother's vaginal secretions, then suckles at her breast shortly afterwards; this constitutes both what is termed "vertical and horizontal transmission." Family bonds through cuddling and kissing constitute horizontal transmission. The baby's health is dependent on this.

C-sections inoculate the baby with unnatural bacterial colonies; they don't get exposed to the same fecal and vaginal flora of the mother—in fact they are exposed to her skin flora and that of other people in the delivery room. Moreover, because the mother is likely to be given antibiotics throughout labor (to prevent group b-strep), the flora is likely diminished also.

After the baby is born it gets exposed to flora from breast skin when it feeds. Feeding causes the consistency of breast milk to change, which then becomes part of the baby's gut microbiome when it suckles. Suckling triggers certain types of immune cells to be activated and deactivates those we don't need. This happens in the first few months of life and educates the immune system to what should we be allergic to and what we shouldn't.

The balance between "innate" immunity versus "adaptive" immunity (*Th1* or *Th2* dominant *T-helper cells*) sets the tone for the reactivity of our system[20] to both internal and external threats for the rest of our lives. The medical community doesn't agree on all the details of how these two types of immune reaction naturally balance [Kidd03] but we do know that natural birth with its inoculation of microbes helps this happen. Microbes play a key part in keeping the body in immune balance via their interactions with the mucosal epithelium (gut "skin" cells) [Bulek10]. It is a fascinatingly complex immune dance that has evolved between us and our microbiome for centuries. We mess around with it at our peril.

Having a C-section is a major determinant of adult disease; if mothers knew this, they might think again about choosing this surgical intervention (i.e., elective where it is not required medically). If it is required, then there are ways to mitigate the negative effects (see later) but it is definitely not without consequences. There has been an incredible increase in this option in Western nations, especially the United States, due no doubt to the higher fee charged by surgeons which rewards it. The result in the child is initially increased atopy (allergies from Th2 dominance) and finally leading to autoimmune conditions (which are a chronic imbalance of innate immunity—Th1), heart disease, and chronic disease generally. The failure to balance properly in the beginning of life means the immune system swings about wildly without the checks and balances of a good microbiome.

Now, if a Caesarian section has to be performed for medical reasons, amelioration of that situation will be necessary. Artificially adding the microflora via a vaginal swab may help, but one needs to check that the mother has a good balanced flora. If not, then both may need to be supplemented. Certainly, it should be done routinely and the shortfall, if present, should be made up as soon as possible with appropriate probiotics.

Thyroid Health

Thyroid health is so important and often missed; a lot of mothers are low in thyroid hormone (hypothyroid) when they give birth. Thyroid hormone is an iodine-containing molecule which helps us produce energy, among other things. People with low thyroid often feel sluggish and low in energy

---

[20] The immune system is divided into two types of response for two different types of threat: for internal threats like viruses, cancer cells, and so on. Th1 provides a proinflammatory response. For external threats like pathogens, the anti-inflammatory Th2 response is required. Each has its own advantages and disadvantages. We need both to be in balance but often they become unnaturally skewed one way or another.

and "joie de vivre." They feel horrible in fact. Recent studies have pointed to an alarming increase in hypothyroidism in western nations, with women more affected than men.[21] Official statistics say that prevalence is around 2% of the UK population but significant figures, like Dr. Thierry Hertoghe, president of the International Hormone Society, a respected physician's organization, suggest the real figure may be 20% to 50% of a standard population.[22] The vast majority of these are women and if they get pregnant, the implications for the child and mother are huge. First, according to Aviva Rom, MD, thyroid hormone affects breast-milk production, energy, mood, and the ability to be present with the child [Rom16]. As I described in my last book *The Scar that Won't Heal* in more detail, a child learns emotional regulation from the interaction with their mother. An absence of connection, termed "attachment" in psychology, has lifelong consequences for the nervous system balance of that child. It makes them hyperreactive to stress, as we have already described.

Breastfeeding

In the UK, we have the lowest levels of breastfeeding in the world; a startling statistic is that only 0.2% of mothers are still breastfeeding by the time their babies are one year old. This compares badly with the rest of Europe and other Western nations and particularly disastrously compared with developing nations—that is, in Africa it's about 98% [Gallagher16]. Mothers have been discouraged from breastfeeding by the availability and convenience of synthetic alternatives ("infant formula"). The World Health Organization (WHO) and Save the Children (a UK charity) have said "the active and aggressive promotion of breast milk substitutes by their manufacturers and distributors continues to be a substantial global barrier to breastfeeding [UNICEF16, para.16]." It is a sobering thought as breastfeeding would save 800,000 lives a year alone.

Also, some women, already highly stressed by the birth experience and lacking support, find it difficult to start or sustain breastfeeding due to poor technique (nipple attachment is too shallow) or mastitis and skin cracking (which may be linked to poor skin microbiomes[23]). Modern lifestyles

---

[21.] Hypothyroid is mostly seen in women between the ages of 40–50 and in women ten times more than men.

[22.] Hypothyroidism is being missed by doctors using outdated reference ranges. According to Dr. Raphael Kellman, most hypothyroidism is due to toxic chemicals that act as endocrine disruptors.

[23.] The mother may have an overgrowth of candida (causing thrush of the nipple) or overgrowth of Staphylococcus aureus, both are a normal part of the microbiome but a problem in excess.

generally make it more difficult to breastfeed naturally, discouraging mothers from being able to do it when the baby needs it when in public.[24]

Antibiotics given to the mother are received through breast milk and so most babies already have high exposure to them (30–40%). It is a good idea therefore to give the mother probiotics in the third trimester; particularly Bifidobacterium and Lactobacillus species—they are then less likely to give birth to atopic babies (babies with skin problems, asthma, etc.). The mother should then continue taking the probiotics up to nine months after birth.

Food has become adulterated and that is true of most children's food too. Avoid soy formula (Dr. William Sears showed it should be mammalian milk, or goats-milk based like Nanny Care if cow's milk is not well tolerated). The problem is we are not always well adapted to drink cow's milk, particularly if we lack the enzymes needed to digest it properly. This is a genetic variation which occurs in some Asian, African, and South American cultures where dairy milk was not common in the diet. Some experts recommend we supplement newborns with vitamin D3 as it optimizes their calcium uptake and hormone health. Also, a compatible probiotic promotes less intestinal gas or colic. Colic may be an immature neurologic reaction to bovine immunoglobulin, so avoid cow's milk in the maternal diet as well as the babies diet' (the reaction may continue for three weeks after stopping). Also consider cranio-sacral therapy and touch work, as they help propulsion activity in the small and large bowel and are generally rebalancing, particularly if the birth has been traumatic.

A note of warning; proton pump inhibitors are sometimes given to babies for colic. This reduces stomach acid which also disrupts the absorption of minerals and vitamins. A much better idea is to give probiotics instead. If you want to give your baby a good start that won't upset them, you can add fermented coconut water (kefir) to a baby's drink early on; they might not like it immediately, but they adapt fairly quickly. They are a delight to raise compared to a distressed or colicky child; they sleep well and are happy when they wake up. There are many simple dietary and behavioral changes that do not take that much effort and various websites and blogs devoted to the subject, so I won't go on here.

[24.] With more mothers working, even when their children are young, and the dearth of places where they can breastfeed in public, make it culturally discouraged, even shameful. Women have been known to be asked to leave public places where they attempted to breastfeed a child—even specialist shops like Mothercare!

Child Development

Postnatally, further inoculation should come from growing up in a household where a child experiences being able to crawl on the floor and puts things in its mouth; this has worked for millennia. But it is under threat. Our environment is subtly changing to one that is more sterile due to overzealous cleaning, a lack of extended families and living with animals (that would have increased the diversity of inoculation), and a cumulative effect of *poor diversity in the rest of the family*. Today, we are more likely to have uniparental inoculation, that is, from the mother only, which is potentially very unbalanced if the mother herself has an unbalanced flora; and thus, imbalance perpetuates down the generation. Not only that, but our microbiome is shared with the microbiome of the house we live in. When we move, we take this with us, and when new people move in it changes again. We inhabit our home with our bacteria. What a strange thought!

It is an indisputably recognized fact that children who are prescribed antibiotics as children are more likely to be obese in later life [Turta16]. Imbalance or overgrowth of certain species has systemic (whole body) effects. The practice of giving children antibiotics for minor infections, like the commonly occurring childhood problem of middle ear infections, makes them worse when the antibiotics wear off because it makes the children more susceptible to more aggressive infections later on. Therefore, it is better to give them vitamin D3 (a variant of vitamin D that is best absorbed) and vitamin A (10000 IUs), which limits swelling and reactivity of mucus membranes, instead of antibiotics for small infections. This helps balance T-reg cells (part of the innate immune system) and thus is anti-inflammatory. You could combine this with a tapering regimen of vitamin C (1000 mg sodium ascorbate beginning with a dose four times a day for two days and diminishing the dose down to once a day over a week. Taper it to promote better immune activity). But, as with all these interventions, these are only to be done with the support of a specialist practitioner. Babies' immune systems and gut cells are easily damaged.

Remember that a well-balanced microbiota influences/regulates appetite and so helps weight control. One of the ways it does this is via a hormone called leptin, which helps promote satiation (fullness) which makes you want less food. It's a similar hormone to insulin which helps promote sugar uptake. If insulin fails to be produced or is not recognized by the body, the result is diabetes (type 1 and type 2 diabetes [T2D] respectively).

T2D is far more common and occurs because of overproduction not underproduction. If there is too much of a release, the hormone receptors in the target tissues become resistant to the hormone and increasingly more of it is needed to get a response. The receptors on the tissue become resistant and lose their ability to respond to the insulin that is present; gradual resistance increases as the disease develops. Both insulin and leptin resistance depend on the gut flora being sub-optimal. With an imbalanced gut, the likelihood of obesity increases and if this starts in childhood the risk of becoming obese in later life is massively increased. We need to inoculate our children if they have not inherited a balanced microbiota.

This imbalance also stops children from absorbing micronutrients and so they get incompletely digested foods leading to leaky gut. These issues likely develop while we are young but may remain undiagnosed as we are more resilient (the body can compensate when young) and so we often don't show symptoms until later in life. However, we are seeing a disproportionate rise in allergies and intolerances in the young, that are perhaps the early warning indicators. Whatever the issue, you can be sure it did not arise in adulthood solely.

In the past, most children would be breast fed for two years, while gradually being introduced to veggies (less calorically dense foods and rich in trace minerals, fiber, and vitamin C). These days we can still introduce vegetables, paying particular attention to colored veggies—especially carrots (containing beta carotene). One thing we must avoid is giving high animal protein—a child's kidneys will not be sufficiently developed to deal with high levels of protein. If we know the child has a particular problem, then bone broths can help to repopulate the gut to heal it. (See the section on Body Ecology). The best bones to make a broth with are lamb or cow bones (chickens are often fed soy, unless it specifies corn on the label). Cooked foods allow more calories to be absorbed, along with fiber, which helps make it not so constipating. Introduce veggies and animal proteins after nine months.

Children who grow up in a cooking culture, (not the UK unfortunately), learn directly that foods need to be combined in certain ways, and they get to smell and see the food being prepared within the kitchen. Their relationship to food develops early and is perceived by the senses before being eaten, which stimulates appetite as well as an interest in food as a cultural/emotional event. Compare this to frozen/ready meals which arrive with little preparation and no skill. Witnessing the preparation of ingredients in the kitchen and their subtle combination, implants the importance of spices and herbs to flavor food. Introducing vibrantly flavored foods when young, stimulates interest in flavor and helps develop taste buds and sensory acuity. Palatable ingredients like ginger, garlic, cumin, turmeric, pesto, and tamari can be introduced gradually as taste buds develop.

Remember, also that the immune system is still developing up to age four, so a vaccine, particularly if it is early or a multiple strain type, can promote dysbiosis. I am not coming down either way on the vaccine debate to avoid controversy; that is a personal decision of the parent based on their received wisdom. But at the very least, the baby should receive probiotics to help ameliorate the effects on the microbiome. These can be administered once they reach age two but should be rotated with different strains and manufacturers. Better still, use raw milk or fermented foods like kefir. Less is more here. As Summer Bock, a US-based health writer states, "there's not a single drug that comes close to the power of little changes in the microbiome." [Bock16]. I couldn't agree more.

New Innovations and Ideas

One of the latest innovations within modern medicine that uses the microbiome is a *fecal transplant* from a person eating a good wholefood diet. These are beginning to be done within the National Health Service in the UK for certain very severe allergies, and so on. However, this new application is far from universal understanding is patchy. In point of fact, it is an old remedy, known to Chinese medicine for thousands of years, but like many of these, it takes modern science to corroborate it before becoming widely accepted. It is a rapidly growing area, and seems to be particularly helpful for people with diseases that other treatments cannot touch, such as recurrent C-difficile infection.[25]

The Importance of the Appendix

The human appendix, a narrow pouch that projects off the caecum in the digestive system, has long been known as a vestigial organ with unknown function. It is largely ignored until it gets inflamed (appendicitis) when it can cause high fever and a lot of pain leading to surgical removal. However, this has always puzzled me, knowing the intelligence of the body, why it would continue to sustain a defunct organ with the energy expenditure that entails.

[25.] Clostridium difficile. See the *BBC Magazine* article here http://www.bbc.co.uk/news/magazine-27503660.

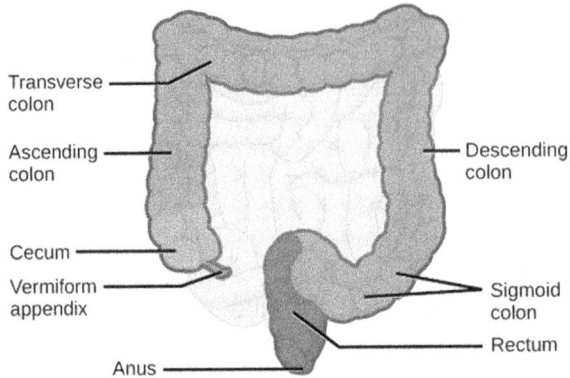

Figure 5.3. The appendix.

Recent research on species has found that species with an appendix have higher average concentrations of lymphoid (immune) tissue in the caecum. This finding suggests that the appendix may play an important role as a secondary immune organ. [Smith17]. Indeed, it may serve as a reservoir for beneficial gut bacteria and removal would therefore be potentially problematic as it would reduce our capacity to keep our gut microbiome balanced. This exciting new research is based on a comparison of different species and is not related to humans alone, but it does have implications for avoiding surgical removal in the future, if this discovery is supported by human studies.

Mental Health—Psychobiotics

The latest research goes beyond merely acknowledging the importance of gut microbes to intestinal health. It also considers the importance of a balanced microbiota for mental health. Experiments in germ-free mice (mice kept sterile so that they have no gut bacteria) often show reduced resilience to stress. The mechanism appears to be via the neuro-endrocrine Hypothalamus-Pituitary-Adrenal (HPA) axis, although no one has yet elucidated how exactly it is mediated, but some exciting new research shows that administering selective microbial species can reverse the anxious behavior of these germ-free mice. Bifidobacterium infantis, for instance, does this. Could inoculation with this species help combat human anxiety? The jury is still out on that but watch this space.

Latest developments have even looked at the autism connection, which has long been known to include a gut component alongside the brain symptoms. One inspiring researcher, Ousseny Zerbo, investigated the epidemiological finding that women who have a prolonged fever during pregnancy were twice as likely to give birth to a child with autism [Zerbo13] unless the fever was controlled. This finding was part of the CHARGE study (Childhood Autism Risks from Genetics and Environment) that sought to investigate the effect of environment on brain development. Altered levels of short-chain fatty acids[26] in the gut may be part of the problem, as may an altered stress response via the HPA axis and neuro-endrocrine hypothalamus-pituitary-adrenal (HPA) axis and autonomic nervous system links.

"Gut microbes are part of the unconscious system regulating social behavior." Germ-free (sterile) mice show autism-like symptoms of reduced sociability, fearful response to noise, and other stimuli. It seems that, to have social interaction, our fear response must be inhibited. This idea has been shown very clearly by Stephen Porges' seminal "polyvagal theory" that shows a hierarchical, three-tier stratification of the autonomic nervous system:

1. Social engagement system (ventral—above diaphragm—parasympathetic)

2. Fight and flight (sympathetic system)

3. Immobilization/freeze (dorsal—below diaphragm—parasympathetic)

When social engagement behavior is properly developed, it inhibits the lower two levels of survival behavior, but if the *social engagement system* (that innervates face, head, and neck) fails to develop properly, the lower levels are activated, and a stress response becomes dominant. This is largely developed from interaction with the mother early in life and will be affected by her microbiome and her stress levels that affect her ability to attune to the child. As the child then develops, it has a heightened response to stress; it seems that microbes are involved in the development of an abnormal stress response where the "set point" (the tolerance level) is set too low.

The microbial presence in the gut is critical to the development of an appropriate stress response later in life. There is a narrow window in early life where colonization must occur to ensure normal development of the HPA axis.

[26.] The components of fats referred to as fatty acids have a number of carbon atoms in the chain that dictates their chemical reactivity and physical properties. A short chain fatty acid (SCFA) has two to six carbon atoms.

This heightened response seems to be reversed with inoculation with certain microbes like B. infantis and Lactobacillus species. So, a new market may open up for artificial inoculation to restore emotional resilience; it is well supported by fMRI studies which show changes in brain activity in response to emotional processing [Tillisch13]. It is a short step to deliberately introducing certain species to treat specific mental health conditions as this indicates.

A recent influential paper has suggested that we should consider promoting mental health through influencing the microbiome as the new field of "psychobiotics" [Sarkar 16]. This promising new approach constitutes a novel class of probiotic with psychotropic (mind altering) properties. According to Dr. Ted Dinan, a psychobiotic approach consists of the introduction of a live organism which is capable of delivering neuro-active substances [Dinan13]. However, the science is still new here; "the mechanisms underlying the ability of stress to modulate the microbiota and also for microbiota to change the set point for stress sensitivity are being unraveled" [Sherwin16]. I refer you to the book by Raphael Kellman *Healthy Gut, Healthy Brain* and his excellent website where he outlines recipes and a protocol for healing depression and anxiety [Kellman17]. There is also the exciting recent research of psychedelic treatments like psilocynin/LSD that help to alter the mood permanently after only one treatment. Psychobiotics may, indeed, be the brave new world of psychiatry, as we realize that microbes don't just help physical ailments but, humans being both and mind and body, mental ones too. And they do this at a considerably lesser cost to the human being than most current psychiatric interventions, which is an exciting development that could lead us out of the current psychiatric pharmaceutical rabbit hole.

How the gut flora influences neurotransmitters in the brain is uncertain but certain ideas are being mooted: via small chain fatty acid (SCFA) fragments in the bloodstream, directly via transmission along the vagus nerve as the main nerve of the brain gut microbiota (BGM) axis and even using tiny strips of genetic code called microRNAs to alter how DNA works in nerve cells. But this is cutting edge research and as yet we don't know.

Already it is hitting the mainstream press as a possible source for new mental health treatments. The potential, given microbes' huge contribution of genes and their rapid adaptation to the environment, seems limitless. It is even being hailed as the "second genome" [Gallagher18]. One researcher Dr. Kirsten Tillich was struck by how malleable the second genome is and how that is in such stark contrast to our own DNA. Could this be the future of medicine? Professor John Cryan has predicted that in the next five years when y go to your doctor for your cholesterol testing, you'll also get your microbiome assessed. The microbiome is the fundamental future of personalized medicine.

References

[Aghaeepour17] Aghaeepour, N., Ganio, E. A., Mcilwain, D., Tsai, A. S., Tingle, M., and VanGassen, S., et al. (2017). "An immune clock of human pregnancy," *Science Immunology* (2017): 2(15). https://www.science.org/doi/10.1126/sciimmunol.aan2946

[Alderman16] Alderman, B. L., Olson, R. L., Brush, C. J., and Shors, T. J. "MAP training: Combining meditation and aerobic exercise reduces depression and ruminate,on while enhancing synchronized brain activity," *Translational Psychiatry* (2016): 6(2), e726.

[Belviranli15] Belviranli, M., and Okudan, N. "Well-known antioxidants and newcomers in sport nutrition: Coenzyme Q10, quercetin, resveratrol, pterostilbene, pycnogenol, and astaxanthin." In M. Lamprecht (Ed.), *Antioxidants in Sport Nutrition* (Chapter 5). CRC Press.

[Bock16] Bock, S. The Microbiome Summit, May 2016. Own notes.

[Blumenthal12] Blumenthal, J. A., Smith, P. J., and Hoffman, B. M. "Is exercise a viable treatment for depression?," *ACSM's Health & Fitness Journal* (2012): 16(4), pp. 14–21.

[Bulek10] Bulek K., Swaidani, S., Aronica, M., and Li, X. "Epithelium: The interplay between innate and Th2 immunity," *Immunology and Cell Biology* (2010): 8(3), pp. 257–268.

[Chevalier12] Chevalier G, et al. Earthing: Health implications of reconnecting the human body to the Earth's surface electrons, *Journal of Environmental Public Health*, (2012): 2012, 291541. https://www.hindawi.com/journals/jeph/2012/291541/

[Cockcroft15] Cockcroft E. J., Williams, C. A., Tomlinson, O. W., Vlachopoulos, D., Jackman, S. R., Armstrong, N., Barker, A. R. "High intensity interval exercise is an effective alternative to moderate intensity exercise for improving glucose tolerance and insulin sensitivity in adolescent boys," *Journal of Science and Medicinein Sport.* (2015): 18(6), pp. 720–724.

[Dinan13] Dinan T. G., Stanton, C., and Cryan, J. F. "Psychobiotics: A novel class of psychotropic," *Biological Psychiatry* (2013): 74(10) pp. 720–726.

[Fawell14] Fawell, J., and ieuwenhuijsen, M. J. "Contaminants in drinking water: Environmental pollution and health." *British Medical Bulletin* (2003): 68(1) pp. 199–208.

[Gallagher16] Gallagher, J. "UK 'world's worst' at breastfeeding," BBC News, January 29, 2016. http://www.bbc.co.uk/news/health-35438049

[Gallagher18] Gallagher, J. "How bacteria are changing your mood.". BBC News, April 24, 2018. http://www.bbc.co.uk/news/health-43815370

[Golan98] Golan, R. *Optimal Wellness*, Wellspring/Ballantine, 1998.

[He16] He F., Li, J., Liu, Z., Chuang, C-C, Yang, W., and Zuo, L. (2016). "Redox mechanism of reactive oxygen species in exercise,"Frontiers in Physiology (2016): 7, article 486.

[Hewings-Martin17] Hewings-Martin, Y. *"What happens to the immune system during pregnancy?"* MedicalNewsToday, September 2, 2017. https://www.medicalnewstoday.com/articles/319257.php

[Hunt12] Hunt, S. J., and Navalta, J. W. "Nitric oxide and the biological cascades underlying increased neurogenesis, enhanced learning ability, and academic ability as an effect of increased bouts of physical activity." *International Journal of Exercise Science*, (2012): 5(3), pp. 245–275.

[Kellman17] Kellman, R. Healthy Gut, Healthy Brain: Heal Depression, Anxiety, and Mental Fog Without Medication By Restoring Your Microbiome. Da Capo Lifelong Books, 2017.

[Kidd03] Kidd P. "Th1/Th2 balance: The hypothesis, its limitations, and implications for health and disease." *Alternative Medicine Review* (2003): 8(3), pp. 223–246.

[Ober14] Ober , C. *Earthing* (2nd ed.). Basic Health Publications, Inc., 2014.

[Oschman15] Oschman, J. L., Chevalier, G., and Brown, R. "The effects of grounding (earthing) on inflammation, the immune response, wound healing, and prevention and treatment of chronic inflammatory and autoimmune diseases," *Journal of Inflammation Research*, (2015): 8, pp. 83–96.

[Rom16] Rom, A. The Microbiome Summit, May 2016. Personal notes.

[Sarkar 16] Sarkar, A., Lehto, S. M., Harty, S., Dinan, T. G., Cryan, J. F., and Burnet, P. W. J. "Psychobiotics and the manipulation of bacteria–gut–brain signals." *Trends in Neurosciences* (2016): 39(11), pp.763–781. http://doi.org/10.1016/j.tins.2016.09.002.

[Sherwin16] Sherwin E, Rea K, Dinan TG, Cryan JF. "A gut (microbiome) feeling about the brain." *Current Opinions Gastroenterology*. (2016): *32*(2), pp.96-102.

[Smith17] Smith, Heather et al., "Morphological evolution of the mammalian cecum and cecal appendix". *Comptes Rendus Palevol* (2017): 16 (1): 39.

[Tillisch13] Tillisch K., Labus, A., Kilpatrick, L., Jiang, Z., Stains, J., Ebrat, B., Guyonett, D., Legrain-Raspaud, S., Trotin, B., Naileboff, E., et al. "Consumption of fermented milk product with probiotic modulates brain activity," *Gastroenterology* (2013): *144*(7) pp. 1394–1401. [Turta16] Turta, O., and Rautava, S. "Antibiotics, obesity and the link to microbes: What are we doing to our children?," *BMC Medicine* (2016): *14*(1): article 57.[UNICEF16] Mason, F., "Failing to breastfeed costs the global economy around US$302 billion every year", The UNICEF, 2016, https://www.unicef.org.uk/baby-friendly/lancet-increasing-breastfeeding-worldwide-prevent-800000-child-deaths-every-year/#:~:text=The%20active%20and%20aggressive%20promotion,BMS%20industry%20is%20growing%20fast.

[Virgin16] Virgin J. J. The Microbiome Summit, November 2016. Own notes.

[Weng14] Weng S-L., Chiu, C-M., Lin, F-M., Huang, W-C., Liang, C., Yang, T., Yang, T-L., Liu, C-Y., Wu, W., Y, et al., "Bacterial communities in semen from men of infertile couples: Metagenomic sequencing reveals relationships of seminal microbiota to semen quality," *PLoS One,* (2014): *9*(10), e110152.

[Zerbo13] Zerbo, O., Iosif, A. M., Walker, C., Ozonoff, S., Hansen, R. L., Hertz-Picciotto, I. "Is maternal influenza or fever during pregnancy associated with autism or developmental delays?" Results from the CHARGE (CHildhood Autism Risks from Genetics and Environment) Study, *Journal of Autism and Developmental Disorders* (2013): *43*(1), pp. 25–33.

CHAPTER 6

DISEASES OF IMBALANCED MICROBIOME

Infectious Disease as Primer of Immunity

We are only now beginning to realize the profound effect your microbiome has on your health and well-being. We are at a significant juncture in our ability to map the microbial genome and see how it interacts with our own to create disease [Cho12]. You will remember that the gut is the first line of defense; "more immune decisions are made daily in the gut than the immune system sees in a lifetime" [Mayer03]. Diseases themselves have an immune-regulatory effect, for example, measles,[1] herpes, etc., allow our immune cells to learn how to deal with more damaging environmental infections. Far from being a disaster, they help to *prime* the system. This would explain why preventing children from getting them is causing more health problems in later life. Recent research has highlighted an association with cancer and infectious disease, with less cancers affecting people who have had measles, for instance.

Some research is speculating that viruses may have been an important stimulus to how our brains developed. Germ theory has characterized them (and bacteria) as universally bad, that is, to be eradicated. But the truth is far more complex, and our current germ theory-based understanding does little to tackle to rising incidence of complex noncommunicable disease (NCD) which constitutes most medical consultations.

---

[1] Measles can be very damaging, however, if it is contracted when very young and the gut is already compromised.

In the next section, we look in more detail at microbiome issues with specific diseases. Some are obvious contenders, as they are seen as primarily "gut issues" like IBS, celiac disease, and so on. But there are others that would not normally be considered a gut problem: arthritis, MS, CFS, to name a few. Most chronic disease is inflammatory in nature. Recently even Alzheimer's disease (AD) was found to be inflammatory in nature, due to excess carbohydrates in the diet (and not enough good fats) making brain cells insulin resistant in the same way as cells of the rest of the body in diabetes. For this reason, it has recently been given the name "type 3 diabetes" to reinforce this link [de la Monte08]. The protein plaques and tangles[2] that were previously thought to be the cause are like cholesterol in the blood; they are there to protect the damage from going further. When you attack cholesterol, you are "attacking the ambulance and not the problem;" according to Dr. Natasha Campbell McBride who has extensively researched this topic.

Cholesterol is in fact an essential molecule in the body [Mercola12]; the precursor for the manufacture of many of our sex hormones and bile acids (important for fat digestion) and an important component of the cell membrane, keeping it flexible and thus allowing it to interact with its neighbors.[3] But it has been unfairly demonized as the enemy, largely because it is extremely profitable to promote the "high cholesterol causes heart disease myth" as it sells statins. These were the number one most prescribed drug in the United States in and number two in the UK.[4] Statins are not benign drugs[5]; they cause major disruption of your metabolism with the significant side effect of muscle pain. They are now even being routinely prescribed to healthy people as preventative, but the science does not support this use for statins; they should be restricted to those who have had a heart attack [Abramson07], if at all. They are particularly dangerous for women as they increase the risk for diabetes, particularly if you are nonwhite. These universal "recommendations *don't distinguish patients by gender*[6] (my italics), and a small, increasingly vocal group of cardiologists

---

[2.] These abnormal deposits show up in brain tissues revealed by scans and dissection staining.

[3.] The cell membrane is like a transducer in electronics, allowing messages in and out via electrochemical signals. Dr. Bruce Lipton in *The Biology of Belief* believes it is the "brains" of the cell rather than the nucleus, as previously believed.

[4.] https://www.telegraph.co.uk/news/2017/12/13/pill-nation-half-us-take-least-one-prescription-drug-daily/

[5.] If taking statins, you must take coenzyme Q10 to help balance the side effects of muscle pain and weakness.

[6.] Most drugs are not tested selectively by gender, so the differing metabolism of men and women as well as dosage is never considered.

believe that's a mistake" [Rabin14, para. 2]. We may be moving toward a medicine that considers gender in prescribing, but it is not here yet [Whitley09]. Women and men have differing hormonal profiles, not just linked to the area of sexual reproduction, but also of fat distribution, blood sugar control, and liver function [Valencak17]. None of that is considered in most drug prescribing. To add to the complication further, there is also the fact of biochemical individuality, which shows us that we don't all process/metabolize drugs in the same way, but that's another book in itself.[7] One size does not fit all here.

What follows is necessarily a brief description of each condition, and it is not intended to be comprehensive. Where possible, I have given suggestions for how to find more detailed information.

Allergies

Allergies are epidemic around the world; mostly in developed nations where diet and lifestyle stress the system and the mitigating effects of a good microbiome are not so prevalent. Generally, allergy is understood as itching, sneezing, and issues involving the skin and the respiratory tract. But allergies are also involved in other issues that seem less obvious like fatigue, muscle pain, and digestive problems.

Basically, there are two main types of allergies categorized by the type of reactions they produce: IgE (type I)—classic allergy and IgG (intolerance). We know that gut bacteria affect intestinal function particularly with respect to leakiness and food allergies. Therefore, a good functional medical doctor will check for GI symptoms: heartburn, burping, diarrhea, and so on as a measure of susceptibility to allergy. You will be asked if you have had antibiotics prior to symptoms—this is a good clue that your gut imbalance may be caused by this. None of the blood tests are definitive. Urine tests can be done to test for gut permeability with foods in which there is uncertainty, but it is always tricky to ascertain which foods are causing what issues without specific food exclusion and reintroduction protocols.

One of the best ways of dealing with the root cause of allergies is to reduce the body burden of overall allergens. The first treatment, therefore, would be to remove all allergenic foods from the diet, for example, soy, dairy, wheat, corn, and so on. If you are not sure what you are allergic to elimination, followed by reintroduction, allows you to monitor reactions more cleanly. This is called the "powerwash" technique. The problem with

---

[7.] Covered in more detail in my previous book, *The Scar that Won't Heal*.

food allergies is the symptoms can be long after the allergen is reacted against and so it can be difficult to pinpoint what is causing it. Removing it from the diet allows you to be much surer, as your body is not subject to multiple allergies. However, it is not easy to do without support, as the diet changes quite radically thus it is best with help from a nutritional therapist. Alternatives are to get IgE and IgG tests, both of which are available online and via nutritionists or supplement suppliers.

Allergy Resolution — ARC: Avoidance, Removal, Colonization

Leo Galland, in his book *The Allergy Solution*, has described a protocol he uses called ARC: avoidance, removal and colonization. He states that inflammation causes the release of nitric oxide, a natural inflammatory chemical, to accumulate in high concentration in tissues. This allows undesirable bacteria like E. coli to proliferate, which itself encourages more inflammation. So, by removing the common allergen inducing foods and then using a suitable herb like berberine (which is both anti-inflammatory and kills parasites and yeasts), you help to "prune the vineyard" or take out the "bad" bacteria leaving the more adaptive ones behind. This is a form of "internal gardening" (a lovely phrase coined by Raphael Kellman) and is very important for your overall health.

Remember, two thirds of your immune system is in the lining of the gut. It naturally has low grade inflammation at a protective basal level. The standard American diet (SAD) creates a high level of inflammation, leaky gut, and metabolic endotoxemia (increase in permeability to the biproducts of bacteria, for example, Lipopolysaccharides (LPS), which causes weight gain and insulin intolerance). In addition, the SAD diet increases the damage caused by pathogenic bacteria like E. coli and B.wadworthii as it reduces the thickness of the protective mucus lining of the gut. Certain bacteria help to balance this destruction; Bifidobacterium infantis is good for immune enhancement as is Lactobacillus plantarum.

Testing

If you test positive for LPS (*lipo-polysaccharide*) in the blood (or the antibodies IgG or IgA to LPS), it is a sign that the gut wall is unhealthy— it is a good test but unusual still in normal medical practice so you would need to go to a functional medicine doctor or nutritional therapist. You can also test for zonulin (which opens up the gaps between gut cells) and occludin levels (e.g., with Cyrex labs) and oxidized LDL but these are likely to be expensive tests.

Autoimmune Disorders and the Gut

The definition of an autoimmune disease (AID) is that your body is attacking itself. The body's immune system mistakenly believes part of its own body is foreign and attacks it. Examples are rheumatoid arthritis (RA), celiac disease (CD), Hashimoto's thyroiditis, systematic lupus erythematosus (SLE), scleroderma, and others. They are increasing in incidence year after year, and remain, in large part, a mystery to medical science. We could say indeed that it is the most significant epidemic of chronic illness in the modern world; it seems to have taken over from infectious disease as the "new frontier" of medicine. From the point of view of the microbiome, it is a bacterial "coding" or information exchange that goes awry. We know the symptomology and we know some of the developmental stages, but the cause is more elusive because it is, as ever, multifactorial.

Alessio Fasano, a leading AID researcher, has shown that there are three components to autoimmune disorders: genetic predisposition, external triggers, and intestinal permeability (leaky gut). Links with the gut and thyroid (which is an important gland of the endocrine system) are very strong. AID begins with the failure of the integrity of the gut barrier, which needs to be operating properly to keep the cellular junctions tight and prevent triggering molecules from being able to pass through to the blood stream and stimulate the immune system. AID can be seen as a trigger for worse diseases too, so it is important to deal with it promptly where it occurs. When you have constant gut leakiness, this creates a potent inflammatory response causing myriad seemingly unrelated symptoms like bloating, gas, muscle pain, rosacea, interstitial cystitis, fibromyalgia, and so on. See the gut barrier in diagrammatic form in Figure 6.1.

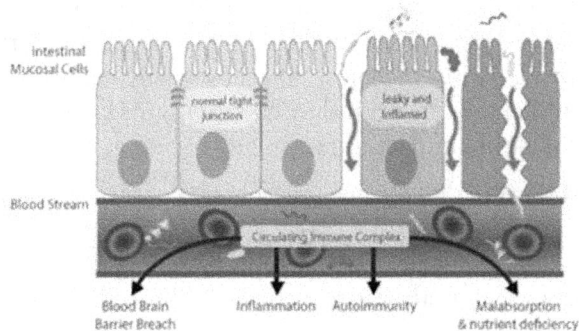

Figure 6.1. The leaky gut.

Steps to Identify Leaky Gut

- Genetic predisposition test: certain genes predispose us to certain AID's. For example, get the HLA sampling genes for celiac. HLA typing helps to identify those likely to develop full blown disease.

- Microbiome related triggers: for example, dysbiosis: stool test should include pancreatic function, short-chain fatty acids (SCFA) profile, calprotectin, lactoferrin, and so on.

- Environmental toxins, other infectious processes: urinary glyphosate testing is starting to be offered with some labs, but with AID it is generally acceptable to presume the patient is toxic. Tests for mercury (Hg) and lead (Pb) via hair, urine, and blood test show whether someone is hanging on to or excreting toxins.

- Pathogens and parasites may be present and will need to be identified is indicated.

- Stress hormones noradrenaline (NA), adrenaline (epinephrine in the United States), and cortisol (measured via salivary test) can alter the microbiome composition and levels can help identify whether chronic stress is a factor.

Treatments

The first step in treatment is via dietary change. This should start straight away without waiting for the results of tests, as it absolutely vital to change the gut barrier permeability which is generally assumed to be involved. Sensitivity to foods, alcohol, drugs, and so on is usually a key indicator and is the result of leaky gut—still, it may be necessary to do an IgG allergy test because it can help identify permeability. Once you heal the gut you lose the allergic response to things—this has certainly been the case for me. This is missed by the vast majority of medical intervention which tries to medicate the symptoms only. For instance, note how many children take inhalers to school—wouldn't it be better if they had their guts healed instead? Perhaps not for the inhaler manufacturers.

The first steps in diet change are a change away from high carbohydrate diets (the majority of western diets) to a high fat, moderate protein, and low-carb diet including, ideally, broths and animal fat (if no moral objection). Both smoothies and broths are a good way of introducing more of the high-fat foods, as they are in a form that is already liquid and contains enzymes, so they are in a sense "predigested." This gives the gut a rest from needing to break down the food and allows you to introduce lots of veggies, berries, nuts, and seeds which would be difficult to eat in the quantities needed. An organic (GMO/glyphosate-free) diet is a must when you have these conditions—particularly if you live in the United States where GMOs are so widespread.

Next, you need to identify and eradicate pathogens—use herbs as a first line treatment. Use marshmallow, plantain, glutamine (in powder form), vitamin D3, vitamin A, zinc, and fish oils as supplements. There are some common parasites associated with specific diseases, for example, Pseudomonas and Clebsiella can be present with Crohn's. For some people, there will be the added problem of small intestinal bacterial overgrowth (SIBO), which is where bacteria that should only be present in the large intestine (colon) migrate into the large intestine and duodenum causing problems. The best way to identify this issue is to have a breath test for hydrogen and methane—if positive, you will need to be prescribed an antibiotic like neomycin or something similar (depending on whether you are hydrogen or methane prevalent). There are also specific supplement formulations for this.

There is a need to balance the gut with good probiotics; Lactobacillus species are best for this. Different species work for different problem types—spore formers like Baccilus coagulans better for high lactic acid producers or people on antibiotics for instance. However, before attempting this yourself in an ad hoc way, I would highly recommend getting advice from a good nutritional therapist or functional medicine doctor.

Anxiety and Depression

It may be strange to think of anxiety and depression as relating to the microbiome, but when you consider that the brain is part of the body, you can see the connections to imbalance and toxicity in the gut and thus the blood and lymph. It may be helpful to think of it like a blocked drain; if the brain is clogged with toxicity, it cannot function properly. This will cause a number of neurological symptoms including nightmares, hallucinations, panic attacks and mood changes.

There are many ways in which you can reverse this, some of which will be relevant to our discussion; you can add serotonin precursors 5HT for serotonin, phenylalanine and tyrosine for dopamine, threonine for GABA. More important is learning how to calm your stress levels with meditation and movement. I believe that anxiety is often due to unexpressed or resolved emotions around a traumatic event in your life, something I discussed in greater detail in my previous book *The Scar that Won't Heal*. So, any nutrient or physical therapy would be best teamed with some calming and nurturing practice too. This notion of "getting in the flow" of your life speaks to the gut flora directly with feel good hormones, neurotransmitters and tiny protein fragments called peptides. When you learn to cherish your gut, you automatically nurture the rest of your body.

Anxiety and depression are not the same disease, but they are very linked under the term "mood disorders" which imply an imbalance of neurotransmitters. However, as we know it is not simply a case of it being a serotonin deficiency. There are reasons why the balance goes awry, which may be down to a priming of the stress response from early (and even prenatal) experience [Mayer15]. A fetus shares the blood supply of the mother so if she is anxious, it is likely those stress hormones will flood the baby's developing nervous system too, with lifelong consequences: a lowering of the "set point" of anxiety.

Taking antidepressants does nothing to cure your depression it just manages it; sometimes this is enough for your brain to then heal itself. To really look at the causes, you have to reconnect with who you are in life and find out what is out of balance. There are likely to be multiple factors: from diet, unresolved emotional stress, and so on, that create anxiety, but the gut is central. If we can't absorb nutrients, this creates an internal stressor that affects the adrenals and raises cortisol, producing an increase in inflammatory chemicals directly affecting our mood as they pass through the blood brain barrier (BBB) into the brain.

Standard medications like benzodiazepines and SSRIs cause huge problems when taken long-term. Anxiety can be a side effect of other medications for insomnia, tooth pain, and so on. The pharmaceutical approach is wrought with problems. But when people get anxious, they reach for the nearest solution, and thus it is the most likely option they will be offered. Your doctor is unlikely to talk to you about nutrition as they do not have any training in it (although that is slowly changing). Certainly, changing your fatty acid balance toward omega-3 rather than omega-6 with supplementation of EPA/DHA via fish or algal oils is a good first step.[8]

A third of all adults are affected by some digestive disorder and it is highly correlated (linked) with anxiety and depression. Much of this is

---

[8.] Most people are now very imbalanced in these—processed foods and meats are high in omega 6, so our balance has been skewed.

modulated by the gut microbiome; "dysregulation of the gut microbiota composition has been identified in a number of psychiatric disorders, including depression" [Sherwin16, p. 96]. Indeed, there is a new scientific understanding developing of the bidirectional communication of the brain-gut-microbiota (BGM) axis that we looked at in Chapter 1. We now turn to how manipulation of the gut flora via diet, probiotics, and prebiotics can influence two-way communication between the three [Liu15] and help to treat most common diseases. Most people who have these conditions have no idea that their gut is involved, and it is seldom discussed in conventional approaches. But it should be a first line approach to *all* disease—as Hippocrates, the "father of medicine" said: "all disease begins in the gut."

Disordered Eating

Bulimia/anorexia and all eating disorders have anxiety (and thus gut dysbiosis) at their base. Basically, with heightened unconscious stress, a person may turn to food to "fix" the problem; either by starving themselves or bingeing and then making themselves throw up or by taking laxatives. This can sometimes be a temporary situation caused by a life situation. If it persists into a chronic condition, there may be brain changes that come from the dopamine (reward) system being constantly triggered by food or abstinence. This can then become entrenched like any addiction, so food, how to obtain or avoid it, becomes the sole focus of living and a psychological disorder ensues.

Looking at it from a psychological viewpoint, an eating disorder is where you are using food as a drug; you start to overcontrol your eating pattern. Whenever you obsess about food, whether about eating or not eating, you use it as a drug. Food addiction is a real issue for many people,[9] although some may not develop full-blown anorexia or bulimia. And not everyone who eats food becomes addicted. So, we need to ask the question "What changes in the brains of people with eating disorders?" It seems it is the neurotransmitter levels and of course these are highly correlated with gut imbalance as most neurotransmitters are made in the gut by the gut bacteria.

For these people that have highly addictive personalities, their brain chemistry is such that they get too much of a GABA, serotonin, and dopamine rush in the brain when they eat certain foods. Also, the gut bacteria change when one eats these foods, so one gets further imbalance

---

[9.] Sugar and wheat are hugely addictive foods. Both are highly refined powders.

as the bacteria themselves create cravings for more of the same. Willpower will not be enough in circumstances like this, and you have to set up a system ahead of time where you plan cooking and shopping, so you are not using the easy option "in the moment."

Forcing someone to eat large amounts when they feel like they want to starve is counterproductive; they have to be gently reeducated to show what food does psychologically[10] and their gut flora rebalanced.

Infections: Chronic Lyme Disease

Much has been talked about Lyme disease, particularly in the United States where it is more widespread and damaging.[11] If you have never heard of it, then prepare to. It seems to be spreading in vulnerable people (i.e., those with an already compromised immune system) but is difficult to identify from symptoms alone as it masquerades as many other more common diseases.

Lyme is a spirochete bacterium called Borrelia, that enters the bloodstream via a tick or mosquito bite (but also from saliva, semen, and vaginal mucus), and then lodges in our "dark places;" either the joints or the gut. It then travels systemically to other regions of the body where it burrows into cells where it cannot be detected (its flagella or tails are buried in between its cell membrane so it evades our immune white cells which use the flagella as a detectable marker). The danger comes when it makes a remarkable transformation; it mutates to a cyst form and inserts its DNA into our own. Once that happens, it can create symptoms that appear to be something else. Conditions it is able to mimic include MS, Crohn's disease, dolitis, CFS, RA, an so on. No wonder it's called "the great pretender." Some consider it a form of "biotoxin."[12]

This is a growing problem resulting from the compromising of our microbiome, which allows the Lyme bacteria to survive and proliferate.

[10.] Psychosensory treatments like tapping (emotional freedom technique [EFT]) and havening are both highly effective for mild eating disorders. For full blown conditions a deeper psychological approach is needed.

[11.] According to expert Dr. Dietrich Klinghardt, the United States and European versions are different, with the US variety being much more virulent.

[12.] There is another theory that it may have become more prevalent as a result of vaccinations for Lyme created by GlaxoSmithKline in 1998. The vaccine was withdrawn due to the fact it created more disease than it solved. But a new vaccine is about to be trialed in the United States and Belgium.

EMFs and GMOs may help it mutate into forms that are more virulent, but a large factor is also the reduction in diversity of the microbiome that would have helped overcome it. Solutions for Lyme have traditionally been large doses of antibiotics which, although they do work, kill off everything and so further contribute to *dysbiosis* (the imbalance of gut bacteria).

Herbs are useful in particular protocols, as are specific Bacteriophages[13] in solution. These phages are viruses that burst open the Borrelia by invading them and hijacking the DNA machinery to make more phages, which then swell the bacterium to bursting point. If we can harness these, we may be able to overturn this alarming chronic infectious disease that is now threatening to overrun us. The problem seems to be that we are amazing incubators for an organism that previously existed in the cold acidic environment of a tick's belly. Being warm-blooded and pH neutral as humans, it has allowed it to become more virulent.[14] There is also some controversy that it became further spread by a vaccine developed against it in the late 1990s, which contained elements of the protein coat that identifies the virus. At the very least this caused some autoimmune problems in susceptible people. The vaccine was subsequently withdrawn after a class action lawsuit brought by a group of people affected.

Treatment

If you suspect Lyme disease you are recommended to contact a specialist naturopathic practitioner as most GPs will not know much about it beyond the classic acute infection type distinguished by a "bullseye" rash. A naturopathic or functional medicine doctor will be needed for chronic Lyme disease. It will require a mixture of antiparasite medications and herbal protocols that start by eliminating toxins that move in size sequence from parasites, mold, Lyme. to viruses.[15] But it needs to be tailored to you individually so cannot be self-treated.

However, one simple change you can make yourself is to eliminate excess sugar and refined carbohydrates from your diet. That cuts out the immune suppressing factors and the food that these bacteria particularly thrive on. Many people have done this just for general health, but it clearly has more specific benefits for this disease.

---

[13] These are viruses that live in bacteria.

[14] See Jack Tips http://www.wellnesswiz.com/.

[15] Dr. Klinghardt's protocol is widely regarded; it usually avoids using antibiotics that are the mainstay of many treatments. He uses herbal compounds such as berberine.

Fibromyalgia and Chronic Pain Syndromes

Fibromyalgia (FM) is a chronic pain syndrome, whereby a person will experience unremitting pain in the postural muscles of the head, neck and back – and they also often have allodynia – pain from even nonpainful stimuli like touch. Sometimes they are in pain just from wearing clothes. This appears be related to the way the brain interprets pain; there is "central amplification" of signals in the brain, so that nonpainful stimuli are perceived as painful. However, that's not the "cause" in a simplistic sense, but part of the chain of reactions related to chronic stress and toxicity; it is a systemic inflammatory and then immune deregulatory illness, but its *origins are in the gut*. Indeed, the high degree of overlap between fibromyalgia and irritable bowel syndrome (IBS) suggests that these conditions are two variations of a common patho-physiological process—small intestinal bacterial overgrowth (SIBO)."

Bacterial debris like lipopolysaccharide (LPS—a form of endotoxin) and other antigens absorbed from the intestine during SIBO, contribute to a subclinical inflammatory state that results in pain hypersensitivity. LPS has been shown to cause increased sensitivity in both animal models and human clinical trials [Vasquez16]. It does this via direct and indirect effects on mitochondrial ATP production: indirectly forming lactate via the diversion into the low energy pathway (glycolysis) and directly via the production of D-lactate by certain bacteria.

In the brain, LPS causes changes in the neurotransmission of the brain cells, particularly astrocytes by overactivating the glutamate receptors and thus causing brain inflammation (a precursor to Alzheimer's) which then exacerbates the mitochondrial dysfunction. It becomes a vicious cycle. In addition, it has just been discovered that cytokines (inflammatory chemicals) produced by gut bacteria can stimulate the vagus nerve to secrete neurotransmitters which cross through the *blood brain barrier* (BBB) and affect things systemically [Breit18]. This, as I discussed in my previous book *The Scar that Won't Heal*, can be instrumental in creating the central (brain) amplification of chronic pain syndromes, whereby pain signals are over-interpreted even in the absence of painful stimuli [Worby15].

Immune dysregulation also triggers too much histamine (from mast cells) in the body (see earlier section on neurotransmitter histamine in Chapter 3). There is also a link to too little HCl acid in the stomach. The gut dysbiosis is severe—but, unlike in IBS, the symptoms tend to appear in the musculoskeletal system rather than just the gut itself. Here, the altered messages contribute to an alteration in energy metabolism in the mitochondria much like in chronic fatigue syndrome. According to Alex Vasquez, a naturopathic physician:

"Treatment of SIBO can be accomplished with berberine, emulsified oregano and combination botanicals, while supplementation with vitamin D and ubiquinone alleviate mitochondrial dysfunction and the central and peripheral contributions to pain, respectively. Most FM cases will respond favorably to this pathophysiology-based approach, while others may require more intensive therapy" [Vasquez16, para. 25].

So much for the conventional approach which ignores the gut altogether and relies on pain killers, anti-epileptics, sleeping pills, and so on. It is my experience that people treated in this way tend to get worse, not better, as the psychological effects of having an "incurable" disease lead to hopelessness which piles on shame to an already hyper-sensitive stress system. In my last book,[16] I talked about how childhood chronic stress (usually of poor attachment to parents and most especially the suppressed rage of betrayal of trust) is highly implicated in fibromyalgia, so I won't go into detail here. Suffice to say it "primes" the stress response to be oversensitive to threat. So, when something stressful happens in adult life, just as in ME/CFS its sister syndrome, the body is unable to absorb the extra stress and goes into a crash. This is usually the point at which people get a diagnosis, although they will have had symptoms of poor immunity, fatigue, and insomnia before that.

So, treating the gut, chronic stress and mitochondrial dysfunction is imperative and needs to be done in tandem (although most people start with the physical functioning first). The psychological components *must be addressed,* however, for lasting change and recovery; we do this by changing the compensatory behaviors of isolation, immobility, and despair. These become understandable entrenched responses to the disbelief of family and friends who can't see anything visibly wrong (when you are fighting for your life, but no one believes you, you will eventually isolate yourself from them to survive). Joining together with other sufferers can be a double bind too, as many are stuck in the belief that their illness is incurable, and they become resigned to disability and managing the disease. The difficulty in finding sources of support where belief in recovery is possible (and where expectations of an "instant fix"

---

16. *The Scar That Won't Heal*, 2015 (revised 2018). Available on Amazon.

are managed appropriately), are one of the many shortfalls of our current health system.[17]

Irritable Bowel Syndrome—IBS

Irritable bowel syndrome is another one of those "mystery" diseases that modern medicine struggles to understand. As we have already said in the section on fibromyalgia and CFS, there is a lot of cross-over with these other "syndromes." This is a medical term for a multifactorial disease for *which no known cause has been identified*—it is shorthand for "we don't know what this is." The UK's NHS approach is one of identifying the dominant symptom (constipation or diarrhea and then treating accordingly, that is, symptom-based). It is not terribly successful, and most sufferers are taught, as with CFS and fibromyalgia, that once you have it, it is for life. The best you can do is "manage symptoms."

However, there is a better *systems approach* which seeks to rebalance the gut flora, reestablish the integrity of the gut lining with diet and supplements, and get off pharmaceuticals as soon as possible. Most people aren't even aware that when they eat conventionally grown/reared food they are unintentionally medicating themselves from the antibiotic/glyphosate residue present in these foods. So, it's not only the pharmaceuticals you are prescribed that cause the problem, but also the foods you eat.

Ideally the first approach should be to repopulate the gut with good bacteria. There are upwards of 10,000 different species in the gut. So how do we know which ones to introduce? Most scientific studies have shown Lactobacillus plantarum as being effective; it is the most studied species and seems to heal the gut lining and thus is able to stabilize immunity, induce free-radical mediated activity (which kills pathogens), activate a metabolic pathway that increases detoxification,[18] and increase antioxidants in your body. However, there are many other species that seem to have beneficial effects such as L. brevis, Bifidobacterium longem, and B. lactis.

A recent meta-analysis study showed that "probiotics that contained Bifidobacterium "appeared to be the most effective" for constipation [Dimidi14] (they balance the immune system and are anti-inflammatory).

[17.] There are organizations who work in a multifactorial way addressing the psychology of the stress response as well as nutrition, exercise, and support; The Chrysalis Effect and Optimum Health Clinic are two in the UK. They are to be encouraged—hopefully one day the NHS will do the same.

[18.] NRF2 for all you science geeks out there!

However, be aware that studies showing detailed protocols for which probiotics to use in what amounts are just not there yet. According to Dr. Allan Walker, director of the Division of Nutrition at Harvard Medical School, there is "still not enough evidence to recommend a specific probiotic to people with constipation" [Corliss 19, para. 6] so the sufferer is left to experiment on their own. Given that probiotics are generally "safe and without side effects, this is probably a safe bet" [para. 8]. However, as always, I would suggest getting a personalized protocol produced for you by a qualified nutritional therapist or functional medicine doctor. Results of such an approach are usually more consistent than a "try it and see" approach.

Myalgic Encephalomyelitis/ Chronic Fatigue Syndrome (ME/CFS)

ME/CFS is a serious chronic condition that affects millions of people in the Western world causing untold misery and dysfunction. The economic consequences are huge. Largely because it doesn't fit the "specialty" model of modern medicine, being a complex, multifactorial condition not limited to one organ or area, it has defied efforts at treatment or even proper diagnosis. Indeed, people who suffered from it were first ridiculed as having an "imaginary" disease that they were making up for personal (psychological) reasons.[19] Clearly this is nonsense. Most of the people who get these conditions are over-functioning, highly productive people, not shirkers as were commonly assumed. But as has become clear, they are seriously ill, albeit with a disease that has defied the standard markers of testing and therefore not fitting the dominant medical model of "one disease/one treatment."

Indeed, as a recent study concluded, "patients with ME/CFS are more functionally impaired than those suffering from type II diabetes (T2D), congestive heart failure, multiple sclerosis (MS), and end-stage renal disease" [Jason09]. It is a *functional metabolic disease* with whole body, systemic effects—many sufferers describe it as a "living death" and, in many ways, this is accurate. The systems of the body start to close down, and function becomes minimal—just enough to keep the person alive but with no quality of life whatsoever. No wonder some reach a point of despair and commit suicide. Most suffer in silence and continue to get what relief they can from ad hoc treatments and painkillers.

[19.] It was first called "yuppie flu"; a derogatory term related to young upwardly mobile executives who started reporting it. However, when you look at the demographics, now it is seen to mostly affect middle-aged women and young adolescents.

However, that is slowly beginning to change. New information is just coming to light which gives hope to sufferers. According to "Mr. Mitochondria" Robert Naviaux, the cell danger response (CDR) is highly implicated in CFS and related syndromes [Naviaux14] such that it can be thought of as a primary failure of the mitochondrion or *mitochondriopathy*. This CDR involves:

- a shift from ATP/ADP metabolism in mitochondria to AMP within the cytosol (cellular fluid)—less efficient.

- a stiffening of the cell membranes to limit super-infection and pathogen egress—to isolate the disease.

- the release of antiviral/antimicrobial chemicals into the pericellular space.

- an increase in mitochondrial fission (autophagy—break up and destruction); and phagy (eating) of intracellular pathogens—less energy production.

- a change in DNA methylation and histone modification to up-regulate pathogenic (disease-causing) gene expression.[20]

- the mobilization of endogenous retroviruses to give genetic variation and warn neighboring cells and effector cells with the release of extracellular nucleotides, peroxide, eicosanoids, cytokines, and more.

It is meant to be a short-term response to limit damage but, if it persists, whole body metabolism and the gut microbiome are dangerously disturbed.

Further problems occur because when the danger is past and cellular function is restored, a metabolic memory of the exposure that led to the CDR is stored in a way similar to the way the brain stores memories, in the form of durable changes in mitochondrial (..) metabolites, cell structure and gene expression via somatic (body-based) epigenetic modifications. This cellular memory is also called "mitocellular hormesis" [Naviaux16] and is under the control of ancient parts of the brain (the brainstem) which, according to a very influential paper in *Science* magazine, may explain the psychological effects [Xie13].

The cell responds in much the same way as the body, by closing down and minimizing wastage of resources and protecting its molecular machinery.

---

20. For instance, pathogenic bacteria methylate (add methyl groups to) mercury in your gut causing all sorts of problems.

Far from being just collateral damage, "the fatigue in ME/CFS, Naviaux believes, is due to *an active and purposeful inhibition of the mitochondria. The mitochondrial are not broken – they're throttled back to a low idle. In the face of danger, they're shutting down and exporting ATP outside the cell to warn other cells that danger is present*" [Johnson16, para. 8].

This is, quite simply, the most important and groundbreaking study of ME/CFS to date because it has identified biochemical markers of a disease (via *metabolomics*[21]) long thought of as imaginary or "all in the head" of the sufferer. It owes its strength to its reproducibility and is a result of "thinking ecologically," rather than simply biologically. For the first time, we can show that CFS has an objectively identifiable chemical signature that may prove useful for both diagnosis and personalized treatment [Johnson16]. However, there are other factors to consider.

People with CFS sometimes have often had a history of long-term antibiotic use, which of course massively affects the microbiome. A single course of antibiotics has been shown to affect the microbiome for up to a year but some studies show it takes two to four years to be restored properly.[22] Taking a course of Saccharomyces boulardii, a natural fungus, concurrent with the broad-spectrum antibiotics, has been shown to be effective in restoring natural diversity. It is safe, even with people who have candida overgrowth. Leaky gut is a huge contributor too, largely brought about by lymphatic dysfunction [Perrin07] which of course affects the brain too, as we are just beginning to discover. In the past, it was believed the brain did not have a lymphatic system. This is being gradually elucidated in exciting new research. See the section on glymphatics for more information. There is significant gut *and* brain inflammation in CFS according to Lakhan [Lakhan10] that has a variety of causes related to dysbiosis and poor intracellular regulation.

What causes the CDR to be switched on? It seems that the reasons are multiple; chemical, physical, or microbial threats (gluten, pathogens, heavy metals, etc.)—there was some controversy in 2007 when it was reported that a virus called X-MRV may have contributed to the disease in susceptible people. However, this was later found to be a red herring,

---

[21.] Metabolomics (study of the cell's metabolites) is a huge area of interest at the moment in the study of disease "signatures," that is, a particular array of metabolites, especially with respect to chronic disease states like CFS/ME

[22.] This may be due to differing sensitivities of people; how diverse and resilient their gut flora was to start with, and the length and strength of the dose of antibiotics.

possibly from contaminated samples, and the researchers concluded that rather than being a one virus—one disease problem, *"the body's response to viruses* may play a larger role than the viruses themselves."* Susceptibility varies between individuals, which is why it is difficult to pinpoint the virus as being the cause of disease. Thereafter, it was strenuously denied and the researchers who proposed the theory vilified. But certainly, there is a consensus that ME/CFS is primarily a disorder of the innate immune system, with systemic cellular effects. We know that the innate immune system is in the gut and is highly susceptible to stress. And those same researchers, led by Dr. Lipkin [Alter12], and subsequently other studies have concluded that genes, environment and timing all conspire to create ME/CFS via epigenetic imprinting — [de Vega17], [Chu19].

One of the major triggers seems to be childhood traumatic/emotional stress causing brain changes in the threat detection system called "limbic kindling," triggering an overactive stress response that causes a whole raft of physiological changes in the body. It affects sleep (where your brain normally clears out its toxins via the lymph system) [Fossey04], metabolism, immunological, hormonal, and brain function changes (mood and cognition). This alteration of the stress response is summarized expertly in a paper by Jankford and others [Jankord08] and described in more detail in *The Scar That Won't Heal* [Worby2015]. It is the main reason why short-term changes in the microbiome don't stick—a stressed body/mind will not create the right conditions for a healthy microbiome. You always have to look at wider lifestyle changes in addition to diet with conditions such as these. Any treatment plan will have to look at the whole person, their sum total life experience and attitudes, diet, lifestyle, and environment.

Covid 19 and Long-Haul COVID (PASC)

The astonishing sudden and devastating COVID-19 pandemic has highlighted many similarities to CFS/ME and long-haul COVID. The breathlessness, fatigue, and general malaise has finally lifted the lid on the devastating consequences of an imbalanced innate immune system. The experts agree that it is not the SARS cov2 virus itself that causes the terrible symptoms, but rather the imbalanced response of the body to this infection.

While initial infection seems to be mostly non-life-threatening for the vast majority of people, in a small minority (darker skins, people over 80, diabetics, those with coronary vascular disease, obesity, and those with kidney disease) seem to be particularly susceptible to the acute respiratory distress syndrome or ARDS. Although initially felt to be a respiratory disease akin to pneumonia, as more and more data has come in, it seems clear it is a "silent" (that is, patients are not aware of it) *hypoxic* event similar to cyanide poisoning whereby the hemoglobin in your red blood cells are poisoned and thus can no longer carry oxygen around the body (hypoxemia [Tobin 20]).

This theory is being mainly promulgated by expert virologist and palliative care professional Dr. Zach Bush in his excellent and comprehensive online presentation on the virome [Bush19] and data analysis by his team. Although unfairly regarded in some trolling circles as a "quack," I think if you are open-minded and critical thinking you will see the value of his unique insight, honed from many years of study and experience. This is ground-breaking information of a depth and breadth that is unsurpassed and, like much that is ahead of its time, it is unnecessarily vilified.[23]

He has a more natural medicine bent, having suffered his own health crisis that took him out of mainstream healthcare and toward a more holistic vision. Sadly, whether you accept his concept of our place in the ecosystem of the earth, you will have likely still had to update your whole attitude to health in the time since COVID-19 first appeared. The sadness is that we did not understand the disease at first, and many people died unnecessarily from incorrect treatment via intubation [Tobin 20].

As our understanding of the correct management expands, we must necessarily turn to preventative measures in order to reduce the burden on healthcare. The vaccines have been developed at breakneck speed in an effort to enable a return to normality, but the role of innate immunity (specifically the gut flora and associated lymph tissue) has been ignored. Vaccines in fact ignore innate immunity altogether, focusing instead on the adaptive antibody response. This is unfortunate as there is much prevention that can be done by raising vitamin D levels in a highly deficient population [Pereira20], reducing exposure NOT via lockdowns but protecting the elderly[24] [Schippers21], and, contrary to media promotion, not using hand gels but ensuring proper handwashing with soap and water.[25]

[23.] It was revolutionary physicist Max Planck who said that "science advances one funeral at a time." In other words, it takes the death of the old guard (and their beliefs) to die before new ones are fully accepted.

[24.] It now appears lockdowns had more negative effects than positive so specific protection for the elderly and at risk is would be better

[25.] Hand gels, while superficially useful in most circumstances, give a false sense of security and tend not to be applied properly. They also contain microbiome destroying chemicals.

Long COVID (also now termed "postacute sequelae of Sars-Cov-2"—PASC) is also emerging as a serious threat, but in different way. Defined as symptoms that do not resolve after several weeks after initial infection [Nabavi20]—it may indeed not resolve for many months with symptoms that have a lot of cross over with chronic fatigue syndrome and fibromyalgia (CFS/ FM): fatigue, malaise, muscle and joint pain, insomnia, and mood changes. The chronic ramifications seem to be unrelated to the severity of the initial symptoms as many who develop it had very mild illness.

The latest figures suggest one in ten people develop it in the UK and it is, like CFS/ME/FM but unlike COVID morbidity, a female-dominated disease of around 5:1 mostly within the 35–69 age group [ONS21]. This is no surprise to me as these are the high stress years for most people; the "years of responsibility" when aging parents can be a concern and careers take over. Women's immune systems are very different, being adapted to carrying a baby, and it may be that viruses interact with them differently.

One plausible theory is that a lack of NAD+, an important biomolecule involved in energy metabolism, may be the cause of long Covid [Miller20] driven by nutritional depletion (particularly of vitamin B3). The body responds by feeding in tryptophan, which is the precursor for serotonin, causing low serotonin and recruitment of mast cells to supply serotonin causing histamine release and all the inflammatory and dysregulatory gut and lung symptoms (it hugely disrupts the microbiome). It is a plausible theory because not only does it fit the clinical picture, but it enables a very low-cost intervention (supply of B3 as nicotinic acid) as a treatment with good results. It remains to be seen how we will respond to the huge burden of long COVID sufferers with long-term symptoms and PTSD from invasive hospital care.

Cancer and the Gut

Cancer is such a politicized arena, I feel daunted even beginning to describe the new information coming out of current research as it threatens much of the current funding and treatment models. Cancer is probably the most feared disease of all and has a cultural taboo that makes it hard to discuss rationally. However, I shall try to treat it with due diligence and hope that you understand that I cannot cover the subject in as much depth as I would like because that would be a book of its own.

There are various theories of why cancer occurs. The dominant one for the last thirty to forty years has been genetic damage causing proliferation of cells which then fail to communicate with each other and grow out of control. This is basically correct, but the rationale is not one of random damage and a genetic lottery as we have been taught. It is in fact a highly coordinated response to perceived environmental threat—and mitochondrial signaling has a large part to play in it. Otto Warburg first identified that, in conditions of poor oxygenation, mitochondria will switch energy production away from the efficient mechanism of *oxidative phosphorylation*, toward low-efficiency *glycolysis* outside of the mitos. However, this isn't a one-way switch, but rather a highly intelligent response toward the survival of the whole. It seems that mitochondria pick up the danger message of threat to the system and shut themselves down *in order to protect the cell from damage.*[26] According to Nunn, "Cancer is clearly associated with changes in mitochondrial dynamics and ultrastructure"; [Nunn2016, para. 17] mitochondria start to gather together to form threads or "guilds." When they do this, their communication and therefore function changes.

The question that should interest us here is how the microbiome is involved in this switching? [Garrett15]. Remember that the gut is the home of the first-line immune system or "innate immunity" (sometimes referred to as Th1/cellular immunity). When it gets unbalanced and dysregulated, it can easily be shifted to Th2 dominance (adaptive/humoral immunity[27]). The immune system gets imbalanced and begins to produce antibodies toward not just extracellular (outside) threats but improperly digested food, or influx of toxins from a leaky gut. We need our microbes to help us digest but with imbalance, the immune system gets confused and dysregulated, and can't fight cancer which needs a strong Th1 cellular response. Thus, we can think of as cancer an "inside job," that is, developing from an intracellular change rather than a random external event. If the body is spending a lot of energy fighting itself believing it is attacking external threats, it has less to use on identifying and destroying errant cancer cells [Kidd03].

Current treatments focus mostly on destroying the errant cells, that is, the tumor, which is regarded as the problem. In order to do this, modern medicine has three arms: removal (surgery), burning (radiation),

---

[26.] When cells have to divide, as is the case with systemic threat and inflammation, the nuclear DNA becomes vulnerable to damage from the free radicals of the ETC, so this has to be switched to the alternative pathway outside the mitos (glycolysis). Without the highly information-rich ATP molecules being produced within the mitos, the cells lose signaling data and fail to regulate their growth. The system becomes dysregulated and tumors result.

[27.] Apologies for including all the terms—different scientific disciplines do not agree on terminology.

and chemical killing of cancer cells (chemotherapy). According to Dr. Jill Carnahan, most chemotherapy drugs were derived from pesticides and are very powerful immune modulators via their interaction with the gut microbiome. A powerful inflammatory response from the microbiome is triggered to enhance immunity to attack cancer cells. But it is a very blunt instrument and kills *all* rapidly dividing cells including hair and gut lining with disastrous results. Chemo is one of the few treatments that could be said to be worse than the disease itself. It causes tremendous suffering.

It is not the place to go into the politics of alternatives to conventional cancer treatment here; there are some more naturopathic ways to heal cancer, but they are unsupported within the current medical system and therefore unregulated; often it is left to the patient to decide which way to go. However, one simple recommendation that I can suggest is to give high-dose broad-spectrum probiotics to people undergoing chemotherapy. It helps to clean up the mess left by the chemo. When the gut flora is reestablished, it begins to talk to the immune system again and produces compounds to heal the gut wall. Research in Japan also points to the benefits of medicinal mushroom supplementation to help the body restore after chemotherapy. [Kono08] Sadly, this is seldom done in current medical practice.

The other thing to note is that this energy production shift to glycolysis (to which the body has switched) needs glucose (hence the name), so uses up vast reserves of fat stores (which are converted to glucose)—hence huge weight loss is often a feature of cancer (cachexia). In a mistaken attempt to build the patient up again, sufferers are often recommended to eat sugar-filled foods. This is a terrible idea, as it simply perpetuates the imbalance—much better would be high fat, low carb (ketogenic) diet, high in essential fats like flax oil, avocados, nuts, and seeds plus high polyphenols from colored vegetables, green tea, and the like. For full details see one of the many excellent research-supported books on the subject [Servan-Schreiber11]. There is so much you can do to help yourself and those you love with this disease. Don't assume the victim role—they are the people who succumb to the disease fastest unfortunately.

Celiac Disease

There has been a huge rise in digestive diseases, of which celiac disease (CD) is perhaps the most serious, as it can kill a person who is undiagnosed and continues to eat gluten. It is a classic autoimmune disease against the protein component of gluten called *gliadin* which gets mistaken for a threat and causes the body to attack its own villi (the small "hairs" of the gut surface lining). It does have a strong genetic component; the HLA carb sampling genes HLA-DQ2 and HLA-DQ8 are implicated, but it doesn't mean they "cause" it. According to the online advisory service Medscape "Approximately 95% of CD patients express HLA-DQ2, and the remaining patients are usually HLA-DQ8 positive. However, the HLA-DQ2 allele is common and is carried by approximately 30% of Caucasian individuals. Thus, HLA-DQ2 or HLA-DQ8 is necessary for disease development but is not sufficient (causative) for disease" [Fasano16, para. 3].

Remember, genes are only a *predisposition*. It takes something in the environment to turn them on. The question is what? I think a dysbiotic gut is central to this answer and this is supported by data. A recent study found that "a reduction in beneficial species and an increase in those potentially pathogenic species as compared to healthy subjects" [Marasco16]. Interestingly, "this dysbiosis is reduced, but might still remain, after a gluten-free diet (GFD)"[para. 1]. Thus, a GFD does not "cure" CD but simply *manages* it. You still need to heal the gut, even if you adhere very strictly to the diet. Ideally this is a low carb, high healthy fats, with fermented foods and bone broths.

Hormone Health—Thyroid Disorders

I include in this general category all the problems of altered levels of the female sexual hormones that cause, variously, dysmenorrhea (painful periods), premenstrual syndrome (PMS), preeclampsia, fibroids, and so on. But more general hormonal regulation is also involved in metabolism (via thyroid hormone), appetite (leptin), and blood sugar regulation (insulin), which affects everyone. Imbalance of these leads to obesity, diabetes, and all the concomitant diseases. These are so common in the population now that for some they are deemed normal. But they are profoundly dysregulated states.

Hormone regulation is also the key to sustained weight loss, however you wish to do it. Otherwise, it will just be a temporary state and then you will put the weight back on. High intensity interval (burst) training has recently been shown to be the most efficient for rebalancing growth hormone and progesterone [Kraemer91] (which is the precursor hormone for most sex hormones including testosterone). Effects seem to be greater in men than women as would perhaps be expected in light of their differing biological roles. But undoubtedly, hormone balance support is a necessary factor in healing most conditions.

Thyroid Disease

One of the most common hormonal issues seen today are issues of the thyroid such as hypothyroidism and its most common form Hashimoto's thyroiditis. The thyroid controls your energy availability so is key to good health. Lack of good thyroid hormone makes someone feel absolutely awful: depressed, low in energy, "brain-fogged," among others symptoms, and it is often missed as a cause of those very common issues. Disease can affect different aspects of the gland.

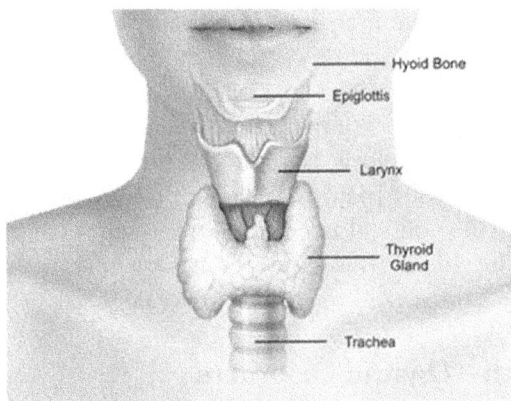

Figure 6.2. Structure of the thyroid gland.

The thyroid gland is a butterfly-shaped tissue found surrounding the trachea in your neck and is part of your endocrine system. It produces a hormone thyroxine (T4) that is converted in the liver, gut, and other organs to the active form tri-iodothyronine (T3).[28] According to Alan Christianson NMD, its importance is often undervalued; dysfunction is extremely common. Latent thyroid disease may be as high as 25% of all adults. The thyroid controls energy for metabolism, so it will slow everything down when you don't have enough thyroid hormone. Particularly, the body cannot get rid of fluid that can damage the heart, so you may often have puffy ankles, wrists, and face. Because it affects all organs, the symptoms are very nonspecific and easily misattributed to age, other lifestyle factors, or diseases. Symptoms include

---

[28.] The naming reflects which was discovered first—thyroxine or T4 has four iodine molecules and was thought to be the only version until T3 with one less iodine was discovered and realized to be the active molecule.

fatigue, weight gain, or inability to lose weight.[29] In addition, the thyroid controls how the liver makes fatty acids (the building blocks of fats) from cholesterol. Therefore, lack of thyroid hormone can cause nonalcoholic fatty liver disease (NAFLD) if there isn't enough to allow the cholesterol to be used properly. Adding to the complication, your levels of thyroid hormone can be either too low (hypothyroid) or too high (hyperthyroid)—the system can go either way.

Lack of thyroid hormones may also be linked to depression, dry hair, and low body temperature. In this case one diagnostic sign is cold hands and feet, and if the body temperature is taken it will be consistently low so the person will most definitely feel colder than other people. As the brain is also a high user of energy, there are often memory problems and brain fog (forgetting words or why you went into a room) and sometimes low mood or depression. Women with hypothyroidism often suffer from bad PMS, and migraines may also be typical. The gut too will have a lot of issues: acid reflux, IBS, constipation, diarrhea, and bloating. But oddly you may not have all or any of these symptoms.[30] Low energy is typical though. If you have intolerance to heat/cold, vertically ridged nails,[31] and occasional tingling and numbness in the extremities, suspect low thyroid.

Dr. Denis Wilson, a US-based thyroid specialist, noticed that low body temperature can also cause anxiety—because the adrenals must pump out a lot more adrenaline to get the blood circulating properly. This is particularly true at nighttime when you fall asleep, your blood pressure drops, and you need to keep a good blood supply to the organs. This adrenaline release is the origin of panic attacks and night terrors—a hypothyroid symptom. He coined this "Wilson's Syndrome," and he promotes a treatment via supplementing thyroid hormone as the active T3 not thyroxine (T4) [Wilson23]. You can see this is a complex area.

With excess T3 so called "hyperthyroidism", you are more likely to get anxiety and palpitations leading to panic attacks; suspect this if the anxiety problems are of new onset and you have many of the other symptoms such as hair loss, muscle weakness, and weight loss.

---

[29] A very diagnostic symptom is a loss of the outer third of the eyebrow.

[30] For me although feeling cold was definitely present, however my main symptom was puffiness and dryness of my eyes.

[31] The nails are actually a very accurate sign, for example, if they have lost their "half-moons" this is a strong indicator of hypothyroid condition.

If unaddressed, then these fairly benign symptoms may develop into more serious problems. Thyroid cancer is the fastest growing cancer in North American women—tripling in the last three decades [ACS23]. European rates are not far behind, with a 74% increase in the UK in the last decade alone [CRCND]. We're not exactly sure why this is; but it is not just due to more diagnosis—there is actually more occurring. I suspect toxicity, low levels of selenium in the diet (which is needed to created thyroxine) and increased chemical exposure. Certain environmental chemicals such as polychlorinated biphenyls (PCBs), found in plastic packaging, displace thyroxine from the carrier protein which allows it to be distributed.[32] Relevant to our discussion in this book is that intestinal dysbiosis (particularly) causes autoimmune thyroiditis (Hashimoto's disease) [Ferrari17]. This is a particularly gut-linked problem of poor nutrient uptake (especially selenium) caused by imbalance of the gut flora.

Testing

Because of the nebulous quality of symptoms, low level thyroid disease is often undiagnosed. Testing relies on measuring various parts of the thyroid hormone (hypothalamus pituitary thyroid—HPT) axis from brain to thyroid gland. The thyroid stimulating hormone (TSH) test (the prohormone from the pituitary) is standard but unreliable; it fluctuates up to 40% on different days (values may be between 1–8).[33] Moreover, the guidelines are lax; the reference ranges need to be updated so that they are lower (ideally between 1–2.5). They were formulated from averages in the population—some of whom are hypothyroid! The other standard test you will receive is blood T4—but this does not tell you about tissue T4 or how much of the active hormone T3 you'll have. So, neither of these tests are much use. If, like me you are dismissed by your GP regarding testing (they won't routinely do a "full panel" thyroid test),[34] you will need to go outside the standard health service to a private functional medicine practitioner or do one yourself online.

Much of the hypothyroid occurrence we see in the population are due to an undiagnosed auto-immune component. We need better testing

---

[32.] Polychlorinated biphenyls, their metabolites and polybrominated diethyl ethers from plastics bind to thyroid transport proteins such as thyroxine-binding globulin (TBG) and trans-thyretin, displace thyroxine, and thus disrupt thyroid function.

[33.] You can also get an ultrasound diagnosis of thyroid disease.

[34.] The justification for this is that it's too expensive. Yet they will allow you a lifetime free prescription for synthetic thyroid hormone (this a standard exemption in the UK).

of antibodies to diagnose autoimmune thyroid disease (See Hashimoto's section)—which isn't routinely done in GP surgeries: thyroglobulin (TGs) and antithyropyroxidase (TPOs) antibodies. According to MD Rachael Wentz, this is because you see elevation in antibodies before changes in TSH (sometimes 10 years before). Thus, we need to treat on this basis rather than TSH or T3/T4 levels which are inaccurate and misleading. The fact that antibodies are present means there is an autoimmune problem (the body is attacking its own thyroid gland) and there is therefore a need to get on thyroid hormone sooner, which helps to prevent full blown disease.

Thyroid/Adrenal Link

Whenever there is a problem with the thyroid you will also need to consider the concomitant adrenal problem. They are very closely linked, according to thyroid hormone expert James Wilson,[35] If thyroid hormone is low, the adrenals then have to pump out too much of their main stress hormone cortisol to get the thyroid to work properly. When this is chronic, we get "adrenal fatigue"[36]—a functional condition not recognized widely by conventional medicine yet, although nutritional therapists are happy to test for this. It is linked to abnormal amounts of cortisol and dehydroepiandrosterone (DHEA), the most abundant circulating steroid hormone in humans. This can be tested for by an adrenal saliva test (done by functional medicine doctors or nutritional therapists). Although there are some labs that you can get to run the test, it is best not to do it on your own as results can be difficult to interpret as cortisol can either be too high or low.

The adrenals and thyroid work closely together and there are many overlaps in function such as fatigue and weight gain. They are both linked to abnormal stress hormones produced throughout the day. Stress is necessary at a low level, it motivates us. The main stress hormone cortisol is anti-inflammatory, so we need some or we get autoimmune disease without it. However, with excess stress, the conversion of thyroid hormone T4 to T3 doesn't happen properly and a warped thyroid molecule reverse T3 (rT3) is produced. This rT3 doesn't activate receptors properly (it is like a dummy key in the lock) so the result is excess pituitary TSH as the pituitary tries to stimulate more production. Hormones operate according to very complex negative feedback loops and with rT3 this hypothalamic pituitary adrenal (HPA) feedback system gets warped into dysregulation as the right feedback is lost.

[35.] Not to be confused with the Dr. Denis Wilson of Wilson's syndrome already mentioned.
[36.] Perhaps better called adrenal dysfunction as the fatigue element is hotly disputed.

Hashimoto's Thyroiditis

This is proper autoimmune thyroiditis and has four stages:

- genetic predisposition

- antibodies

- destruction of thyroid gland

- progression into autoimmune disease (AID) that can go into other organs; such as lupus, RA, among others

Treating Hashimoto's is a systemic problem caused by an impaired ability to handle stress, get rid of toxins, and dysbiosis in the microbiome leading to nutrient deficiencies, leaky gut, and chronic hidden infection. So how do we reverse this problem?

- Start with gluten free, dairy free, soya free diet. This on its own can resolve a lot of problems.

- Introduce a more nutrient dense diet (paleo type) for example, grass-fed meats, organic fruit, adding fermented food into diet. Also, high-fat, low-carb diets can help to balance blood sugar.

- Overturn deficiencies (vitamin D, B12, selenium). Check ferritin levels (if low suggests iron deficiency but can be as a result of not enough stomach acid (hypochlorhydria[37]) too—in which case add enzymes; pepsin and betaine HCl in order to help digest foods. This method is very good for turning around fatigue.

- Detoxification—get tested for heavy metal toxicities.

You can also get a reverse osmosis drinking filter to reduce the heavy metals in the water you drink (or use a charcoal filter but you may need additional mineral electrolytes if you do). You can support the liver with various herbs (milk thistle is particularly good at this). Finally, keep up with your dental health—check for bleeding gums. Work with a holistic dentist if you can as they have more of an understanding of the links between diet and health

---

[37.] Low stomach acid is much more common than high. Drink hot lemon water or a teaspoon of apple cider vinegar in water as soon as you wake up. The stomach is the first line of defense to microbes, so the right pH is important.

Graves' Disease

In this type of thyroid disease, a different part of thyroid gland is attacked, and you get an overabundance of thyroid hormone—hyperthyroidism. This can cause weight loss (loss of muscle mass), palpitations, sweating, and hair loss. Graves' disease tends to have a much quicker onset than hypothyroidism—and the effects are often more noticeable with bulging eyes a good indicator.

It is not a good idea to remove the thyroid gland (as is commonly done with Graves' disease and hyperthyroidism). You often then go on to develop AID as the immune system is still out of balance. You are "killing the messenger" rather than dealing with the problem. Again, look to balancing the microbiome in the gut—get a functional medicine approach rather than resorting to irreversible surgery.

Neurodegenerative Disorders

Neurodegenerative disorders are primarily a sign of autoimmunity. Most of the neurotransmitters (NT) like serotonin (the happy NT), GABA (the balancer), and dopamine (the motivator) are mostly made in the gut. If the microbiome is out of balance, all your neurotransmitters will also be out of balance. So many neurological disorders are primarily digestive disorders; a person who is deficient in neurotransmitters needs gut rebalancing, which is a very different approach to conventional medicine which always deals with the end organ, that is, the brain as the source of the problem.

While there are many diseases related to gut issues (some would say all) that I cannot cover here, it is worth noting that even so-called "incurable" disorders can be turned around by curing the gut issues.[38] This illustrates the naturopathic (natural medicine) idea that all diseases are manifestations of a different "fault line," but the underlying microbial problem is similar so can be healed in similar ways. The exact approach will be customized to suit the person (not the disease—that is where naturopathy differs from conventional medicine). These holistic approaches of nutritional support, osteopathy, dental health, and so on, have allowed complete recoveries to occur. I cannot say that this is possible in every case, but I do know many people who have overcome diabetes and chronic fatigue with similar approaches so it should not seem so outrageous. It is certainly worth trying.

[38.] The film "Documenting Hope" has demonstrated the possibility of fourteen children with a variety of different disorders including autism, diabetes, and obesity being brought through to recovery.

It has been said that changes in the microbiome occur at least five years before diagnosis. This is largely from toxic build up in the brain. There are of course genetic predispositions, that is, a genetic mutation can cause a build-up of particular metabolites in the brain. For instance, in Parkinson's disease there is a genetic susceptibility in dopamine metabolism [Goldstein13]. But does it come from the brain or the gut? Certainly, an altered microbiome is part of the web of disease causation. Here, I am going to outline some of the neurodegenerative diseases that have the best research-supported recoveries from healing the microbiome.

Alzheimer's and Dementia

Alzheimer's disease (AD) is, perhaps secondary to cancer, our most feared disease. Due to its progressive loss of function and deterioration of memory and personality, it seems like a slow death. There is also no known "cure" conventionally.

Alzheimer's has been recently determined as "type III diabetes' (T3D) in that just like its cousin type2D, it is caused by insulin resistance due to an "overcarbified" diet. With an altered glucose metabolism, brain cells become less sensitive to insulin as much as those in the body do. The brain is a metabolically hungry organ, which primarily uses glucose as fuel.[39] Insulin is pumped out of the pancreas when we eat sugar to regulate blood sugar levels by promoting glucose uptake. But if this system is overworked, the blood sugar dips too much and drives the need to eat again to regulate, forcing more and more insulin out of the pancreas to re-regulate it. This can then be like a drug in that the more we become desensitized to insulin's effects, the more we feel we need to consume sugar. In fact, to improve our health we have to train ourselves to restrict sugar and carbs (which act as sugar in the body).

There is a special feature of the brain called the "blood-brain barrier" (BBB), and it has the function of keeping out toxins from the vital processes of the brain. But with a poor microbiome, it becomes leaky just like the gut barrier, so the brain becomes bathed in higher levels of glucose than is good for it. Hence, the best solution, in addition to healing the gut of course, is restricting your sugar intake, that is, more eating of *slow releasing* carbohydrates and more protein/fat. This is the single most important preventative action you can take in your middle age years— before disease strikes you. But there are other interventions that may help

---

[39.] Although ketones have now been found to also be a preferential fuel on a low-carb diet.

us identify and manage this disease. One of these measures is linked to thyroid function.

We see a particular profile with memory loss related to dementia: low thyroid function or hypothyroidism—particularly low thyroxine (T3), high reverse T3 (rT3—an altered molecule that does not fit the receptor properly), and elevation of the inflammatory markers (cytokines) TNF alpha and interleukin 6 (IL-6). Get these tested if you are concerned. Low thyroid is a problem in many diseases, not least cancer, particularly breast cancer so it makes sense to find out your levels.

New types of MRI/EEG are so sophisticated that they can measure levels of glutathione in the brain; a protective, antioxidant chemical that scavenges free radicals throughout the body (and brain). Recent studies have shown this is also raised in AD. This raises hope for earlier detection as glutathione might be considered a possible biomarker for AD [Graff-Radford13]. In other words, the presence of low values of glutathione might be considered an early warning sign of high AD risk. However, as usual, there is much hyperbole around the issue and the politics of how to increase your glutathione have even raged around increased dairy in the diet, with some people recommending this. I personally would question that finding because milk, on the whole, is not a good food anymore given its production and adulteration with hormones. You are better to supplement with lipolyzed glutathione (which is absorbed a lot better).

Turmeric (curcumin) has been shown to profoundly alter the development of the disease for instance, reversing some symptoms to the point that patients began to recognize their family again after a year of taking it. However, the studies are in low numbers and contradictory due to limited bioavailability when ingested (pepper and oil increase its absorption) so you won't see turmeric therapy being offered in your GP clinic anytime soon. The trials are simply not supported as they are nonpatentable since they are foods, and therefore no pharmaceutical companies will fund the very expensive trials. But read the literature and make up your own mind [Hishikawa12], [Mishra08].

According to functional medical doctor, Raphael Kellman, there is no human being that cannot be helped during any stage of the disease; they may not regain their memory, but its progress can be arrested, and, with the help of education, they can be given back some functions. I tend to agree. If, as we now know, AD begins early in middle age as a whole-body response to poor sugar metabolism with concomitant changes in mitochondrial function, then we know it is reversible.

Prader Willi Syndrome

This is an unusual genetic disease whereby people lose the ability to regulate their appetite (via hormones leptin and grehlin) and become compulsive eaters, hence obese and generally depressed (due to lack of serotonin). However, as we are now beginning to realize with all forms of obesity, it is the *epigenetics* that really make the difference. We need to look at why the genes are turned on in these predisposed individuals. An altered stress response may be the culprit with childhood stress and in particular failures of attachment and attunement (consistent mother child bonding) switching on these genes. Children with this condition often have trouble making eye contact and have huge gut imbalances, suggesting a link with the improper coordination of the social engagement system which then has systemic effects.

Making changes to the microbiome significantly reduces the resulting obesity, which clearly shows that it is not only genetics. In a fascinating study in Shanghai, Liping Zhao used dietary change and Chinese herbs to make this change [Ebio15]. These changes were effective because they attack the microbiome imbalances first, which alters the bacterial psychotropic molecules to encourage the eating of healing foods. Dr. Zhao looked at the coordinated groups (or guilds) of microorganisms within the ecological configuration of the gut. By altering the balance so that pathogenic ones are marginalized, he altered disturbances in tryptophan metabolism whereby we don't get enough serotonin (the feel-good hormone that is produced after eating and is responsible for good mood). This illustrates a more systems-based approach to testing and intervention. See the *Nature* magazine article on the international microbiome initiative for more information [Zhao10].

Developmental Disorders

Autism and Autistic Spectrum Disorders

We are currently seeing a rise in autism in children as never seen before. Prevalence has grown to almost one in forty-five children now—a huge increase! This is a very worrying trend as it makes it difficult for children to function well educationally, and also, perhaps more significantly, *relationally* and thus causes great distress to the families. It has huge costs for society at large.

We used to think that autism was caused by bad mothering(!),[40] and latterly it was thought to be the result of a genetic disorder of the brain. It is

---

[40] Unbelievably, this was still taught to medical students as late as the 1970s as the approach of the behavioral school of psychology was dominant.

becoming clear, however, that autism is neither of these things, but rather a combination of extreme toxic stresses on the brain's metabolism caused by various factors of nutritional deficiencies, poor microbiome balance, genetic sensitivities, and mitochondrial dysfunction [Hyman08]. Another multifactorial disease, no less.

There is considerable immune dysregulation that contributes to poor brain development and is possibly linked to C-section births and triggered by childhood stresses at some point in the child's development. The stress of the mother during pregnancy will alter the microbiome balance as we have already discussed. Remember, pregnancy must dampen down the immune system via the innate Th1 system in order for the baby not to be rejected as "foreign" tissue. This results in an increasingly Th2-weighted system, which is much more likely to lead to allergy. Normally at birth this is reversed back to more Th1. If that doesn't happen properly because of microbiota and attunement problems, you get a permanently skewed immune system, which is Th2 dominant. Attunement with the mother, which helps the baby's social engagement system to come online properly; is not only important for immune function but for overall brain development too. If this is disrupted, it is much more difficult to put it right later on.

The pattern that usually forms is that children develop normally for the first eighteen months and then, shortly afterward, they first develop autistic behaviors. This may be triggered by an assault in early childhood, for example, vaccines or antibiotic administration, which can further disrupt immune regulation if it is already stressed. There is a strong genetic component, with contributions from single nucleotide polymorphisms (SNPs—a change in one of the "letters" of the genetic code), which affect the readout of DNA readout increasing symptom severity [Jiao12]. Truly, it is multifactorial and subject to a variety of epigenetic controls,[41] which is why it is difficult to prove one definitive cause. It is, therefore, a hugely controversial subject, like cancer, where arguments rage back and forth. I would recommend you look at the work of Martha Herbert for a good summary; her recent book *The Autism Revolution* [Herbert12] outlines the more holistic vision. She has also written a more scholarly article for the scientific community describing the link to electromagnetic fields (EMFs), too [Herbert13]. She is in no way a "quack," as proponents of

---

[41.] Epigenetic modification of DNA readout occurs via methylation; it is linked to conversion of folate via the MTHFR pathway and if conversion is poor this affects metabolism also.

holistic medicine are apt to be described, but a highly respected medical doctor and researcher from Harvard University. So again, not so easily dismissed.

Julie Matthews, another expert researcher in autism, has shown certain GI symptoms linked to autism. She has identified significant differences in the microbiome between autistic and normal children; 22%–70% show GI distress concomitant with the neurological symptoms. This includes more constipation, pain, etc. While no one is yet saying that dysbiosis is the sole cause, there is a high proportional correlation of GI symptoms with severity of autistic behavior [Adams11], [Krajmalnik-Brown15]. Usually, this proportional relationship makes it a good candidate for something being highly causative. Interestingly, severe self-injurious behavior that was passed off as typical "autistic behavior,"stopped when treated with appropriate medications. It seems that children may be in such severe pain that they are trying to divert or address the pain. The problem is as many can't talk; they can't tell you.

Looking more closely at some of these correlations, there is an increase in pathogenic bacteria, clostridia for instance, particularly for late onset autism. Suterella species are also high as well as desulphovibrio sulphuricans (found in an aquatic environment) that produces sulphide gas. All overgrow as a result of the damage to the immune system which would normally keep these bacteria at bay. Conversely, those affected also have less beneficial bacteria such as Bifidobacterium (reduces inflammation), lactobacillus— possibly linked to a decrease in seafood consumption and less Prevotella species (carbohydrate digesters); this may be why they have a difficulty with carbohydrate.

So, it may be true to say that autism starts in the gut, but it quickly becomes systemic. Poor carbohydrate digesting ability leads to more immune reactivity, more inflammation, less digestive capacity, more pathogenic bacteria, gas, bloating, and pain. SIBO may then develop; these are usually good bacteria which have migrated to the wrong place, that is, higher up the gut where they cause symptoms of bloating and gas.[42] David Hyman has described it as the "hologram" of chronic disease, meaning it has all the symptoms of other chronic diseases contained in it. Certainly, one needs to look holistically (holographically) in order to treat it.

[42.] A test marker for SIBO is levels of D-lactate in the urine.

I recently witnessed the most amazing transformation in an autistic adult with the use of a computer-altered sound program, which was played through headphones and enhanced the prosody (variability) of the human voice to bring online the social engagement system. The originator of this amazing treatment, Stephen Porges, is a highly respected psychiatrist and researcher and has undeniably already created a lifelong legacy with his polyvagal theory. He was reluctant to show us the video of the transformation achievable with just six sessions of listening to his enhanced sounds, because he feared accusations of quackery. But the changes in eye contact, listening ability, facial expression, and ability to enjoy and partake of life were so obvious, they cannot be denied [Porges14]. The man's parents were very clear in their assessment of vast improvement. Moreover, many others have reproduced the effects with children; see the evidence for yourself on their specialist website.[43] Some parents have found their children are being taken off the autism spectrum disorders list. This is modern science at its best, working for humanity without dogma.

If you have doubts, as naturally many of you will have, that sound can do this, watch the recent film "Life, Animated,"[44] in which an autistic boy recovers many of his social abilities through the accidental discovery by his family of the power of accentuated prosody (which directly links the gut, brain, and heart via the ventral vagus nerve). The family mimic his favorite Disney cartoons in everything they say—and their highly autistic child starts to talk again and eventually becomes a man who can now travel the world talking and educating people about autism. He is not "cured" but he is able to function and achieve a university degree and a high quality of life. They were unintentionally using the power of the social engagement system (SES—a branch of the parasympathetic nervous system) to retune their son's brain and reverse his auditory and visual sensitivities. What an incredible true story.

Treatment Approaches

The approach to rebalancing the flora needs to be staged and unique for each person. It usually involves clearing pathogens and then introducing fermented foods. However, a lab test can show markers as to the health of the microbiome so it can be tracked.[45] There are a variety of nutritional

---

[43.] Check out his technique on the listening program website: http://integratedlistening.com/ssp-safe-sound-protocol/.

[44.] http://www.lifeanimateddoc.com/

[45.] Genova labs do one.

protocols, but not all work the same for everyone. However, there are some guidelines:

- Use herbs to treat pathogenic bacteria, for example, clostridia.

- Adopt a gluten- and casein-free diet as a first-line diet intervention.

- Try a grain-free or GAPS diet for those for whom a gluten- or casein-free diet doesn't work. No rice, quinoa, or potatoes.

- Then start a Paleo diet for maintenance.

Some other more controversial approaches include:

- Adopt a reduced-fiber diet like a low FODMAPS (probably best considered after other approaches have been proven not to work).

- Try a low-phenol diet (found in apples, grapes, berries, and raisins). These are normally detoxified by the sulphation pathway. Pathogenic bacteria give off excess phenols and thus this depletes this pathway.

- Choose a low-oxalate diet. Oxalobacter formigenesis is responsible for breaking down oxalate (in fact, it is used for breaking down kidney stones) so it doesn't get absorbed into the bloodstream. Antibiotics can wipe it out. Oxalate interferes with mitochondrial function, so this is a real problem for many common diseases.

Whatever you do, don't assume your condition, even if considered genetic or developmental, is incurable. This modern world instills in us a sense of powerlessness and we need to take it back. In the final chapter, I am going to look now at the paradoxes and vested interests of modern life in more detail.

References

[Abramson07] Abramson J., and Wright, J. M. "Are lipid-lowering guidelines evidence-based?" *Lancet* (2007): 369(9557), pp. 168–169.

[ACS23] "Key statistics for thyroid cancer," American Cancer Society, January 18, 2023. https://www.cancer.org/cancer/thyroid-cancer/about/key-statistics.html

[Adams11] Adams, J., Johansen, L. J., Powell, L. D., Quig, D., and Rubin, R. A. "Gastrointestinal flora and gastrointestinal status in children with autism—comparisons to typical children and correlation with autism severity". *BMC Gastroenterology* (2011): *11*, article 22.

[Alter12] Alter H. J., et al., "A multicenter blinded analysis indicates no association between chronic fatigue syndrome/myalgic encephalomyelitis and either xenotropic murine leukemia virus-related virus or polytropic murine leukemia virus," *American Society for Microbiology* (2012): *3*(5), e00266-12.

[Breit18] Breit S, Kupferberg A, Rogler G, Hasler G. "Vagus Nerve as Modulator of the Brain-Gut Axis in Psychiatric and Inflammatory Disorders". *Frontiers in Psychiatry.* (2018): *13*;9, p.44.

[Bush19] Bush, Z. "The virome: Template for a regenerative future." ZachBush-MD. https://zachbushmd.com/virome-replay/

[Callaway12] Callaway, E. "The scientist who put the nail in XMRV's coffin." *Nature Research* (2012). https://doi.org/10.1038/nature.2012.11444

[Cho12] Cho, I., & Blaser, M. J. "The human microbiome: At the interface of health and disease," *Nature Reviews Genetics* (2012): *13*(4), pp. 260–270.

[Chu19] Chu L, Valencia IJ, Garvert DW, Montoya JG. "Onset patterns and course of myalgic encephalomyelitis/chronic fatigue syndrome." *Frontiers in Pediatrics* (2019): *7*(12).

[Corliss19] Corliss, J. "Probiotics may ease constipation," Harvard Health [blog], June 24, 2019. https://www.health.harvard.edu/blog/probiotics-may-ease-constipation-201408217377

[CRCND] "Thyroid cancer statistics." Cancer Research UK, (no date). http://www.cancerresearchuk.org/health-professional/cancer-statistics/statistics-by-cancer-type/thyroid-cancer#heading-Zero

[Dimidi14] Dimidi E., Christodoulides, S., Fragkos, K., Scott, S. M., and Whelan, K. "The effect of probiotics on functional constipation in adults: A systematic review and meta-analysis of randomized controlled trials," American Journal of Clinical Nutrition (2014): *100*(4), pp. 1075–1084.

[de la Monte08] de la Monte, S. M., & Wands, J. R. "Alzheimer's disease is type 3 diabetes—Evidence Reviewed." *Journal of Diabetes Science and Technology (Online)* 2008: *2*(6): pp. 1101–1113.

[deVega17] de Vega WC, McGowan PO. "The epigenetic landscape of myalgic encephalomyelitis/chronic fatigue syndrome: Deciphering complex phenotypes." Epigenomics. (2017): *9*(11), pp.1337-1340.

[Ebio15] "Getting healthier through microbiome makeover," [Editorial] *EBio Medicine* (2015): *2*(8), p. 771.

[Fasano16] Fasano, A. "Genetics of celiac disease (Sprue)." Medscape, November 7, 2019. http://emedicine.medscape.com/article/1790189-overview, 2016.

[Ferrari17] Ferrari S. M., Felahi, P., Antonelli, A., and Benvenga, S. "Environmental issues in thyroid diseases," *Frontiers in Endocrinology* (2017): 8, 50.

[Fossey04] Fossey M. "Sleep quality and psychological adjustment in chronic fatigue syndrome." *Journal of Behavioral Medicine* (2004): 27(6), pp. 581–605.

[Garrett15] Garrett W. S. "Cancer and the microbiota," *Science* (2015): 348(6230), pp. 80–86.

[Goldstein13] Goldstein D. S., Sullivan P., Holmes, C., Miller, G. W., Alter, S., Strong, R., Mash, D. C., Kopin, I. J., and Sharabi, Y. "Determinants of buildup of the toxic dopamine metabolite DOPAL in Parkinson's disease," *Journal of Neurochemistry* (2013): 126(5), pp. 591–603.

[Graff-Radford13] Graff-Radford, J., and Kantarci, K. "Magnetic resonance spectroscopy in Alzheimer's disease," *Neuropsychiatric Disease and Treatment* (2013): 9, pp. 687–696.

[Herbert12] Herbert, M. *"The Autism Revolution: Whole Body Strategies for Making Life All It Can Be."* Harvard Health Publications/Random House, 2012.

[Herbert13] Herbert, M. R., and Sage, C. "Autism and EMF plausibility of a pathophysiological link— Part I." *Pathopysiology* (2013): 20(3), pp. 191–209. http://www.ncbi.nlm.nih.gov/pubmed/24095003

[Hishikawa12] Hishikawa, N., Yoriko, T., Yoshinobu, A., Yuhei, T., Yoshitake, T., Hisayoshi N., Nobuyuki, M., and Krishna, U. K. "Effects of turmeric on Alzheimer's disease with behavioral and psychological symptoms of dementia,"*Ayu* (2012): 33(4), pp. 499–504.

[Hyman08] Hyman, M. "Autism: Is it all in the head?," [Editorial] *Alternative Therapies in Health and Medicine* (2008): 14(6), pp. 12–18. http://drhyman.com/downloads/Autism.pdf

[Jankord08] Jankord R., & Herman, J. P. (2008). "Limbic regulation of hypothalamo-pituitary-adrenocortical function during acute and chronic stress," *Annals of the New York Academy of Sciences* (2008): 48, pp. 64–73.

[Jiao12] Jiao, Y. et al., "Single nucleotide polymorphisms predict symptom severity of autism spectrum disorder," *Journal of Autism and Developmental Disorders* (2012): 42(6), pp. 971–983.

[Jason09] Jason, L. A. "Kindling and oxidative stress as contributors to myalgic encephalomyelitis/chronic fatigue syndrome." *Journal of Behavioral and Neuroscience Research* (2009): 7(2), pp. 1–17.

[Johnson16] Johnson, C. "The core problem in chronic fatigue syndrome identified? Naviaux's metabolomics study breaks fresh ground." HealthRising, September 1, 2016. http://www.healthrising.org/blog/2016/09/01/metabolomics-naviaux-chronic-fatigue-syndrome-core-problem/, 2016.

[Kidd03] Kidd, P. "Th1/Th2 balance: The hypothesis, its limitations, and implications for health and disease," *Alternative Medicine Review* (2003): 8(3), pp. 223–246.

[Kono08] Kono K, Kawaguchi Y, Mizukami Y, et al.: Protein-bound polysaccharide K partially prevents apoptosis of circulating T cells induced by anti-cancer drug S-1 in patients with gastric cancer. Oncology 74 (3-4): 143-9", 2008

[Kraemer91] Kraemer, W. J., Gordon, S. E., Fleck, S. J., Marchitelli, L. J., Mello, R., Dziadoes, J. E., Friedl, K., Harman, E., Maresh, C., and Fry, A. C. "Endogenous anabolic hormonal and growth factor responses to heavy resistance exercise in males and females," . *International Journal of Sports Medicine* (1991): 12(2), pp. 228–235. https://www.thieme-connect.de/products/ejournals/abstract/10.1055/s-2007-1024673

[Krajmalnik-Brown15] Krajmalnik-Brown, R., Lozupone, C., Kang, D-W, and Adams, J. B. "Gut bacteria in children with autism spectrum disorders: Challenges and promise of studying how a complex community influences a complex disease," *Microbial Ecology in Health and Disease* (2015): 12(26), article 26914.

[Liu15] Liu X., Cao, S., Zhang, X. "Modulation of gut microbiota-brain-axis by probiotics, probiotics, and diet," *Journal of Agricultural and Food Chemistry* (2015): 63(36), pp. 7885–7895.

[Lakhan10] Lakhan S.E. "Gut inflammation in chronic fatigue syndrome," *Nutrition & Metabolism* (2010): 7(79). https://nutritionandmetabolism.biomedcentral.com/articles/10.1186/1743-7075-7-79

[Marasco16] Marasco, G., Di Biase, A. R., Schiumerini, R., Eusebi, L. H., Iughetti, L., Ravaioli, F., Scaioli, E., Colecchia, A., and Festi, D. "Gut microbiota and celiac disease," *Digestive Diseases and Sciences* (2016): 61(6), pp. 1461–1472.

[Mayer03] Mayer L. "Mucosal immunity," *Pediatrics* (2003): 111(6), pp. 1595–1600.

[Mayer15] Mayer, E. A., Tillisch, K., and Gupta, A. "Gut/brain axis and the microbiota," *The Journal of Clinical Investigation* (2015): 125(3), pp. 926–938.

[Mercola12] Mercola, J. "The Low-down on Cholesterol: Why You Need It – and the Real Methods to Get Your Levels Right" https://www.mercola.com/eb-ook/how-to-lower-cholesterol.aspx

[Miller20] Miller, R., Wentzel, A. R., Richards, G. A. "COVID-19: NAD+ deficiency may predispose the aged, obese and type2 diabetics to mortality through its effect on SIRT1 activity," *Medical Hypotheses* (2020): *144*, 110044.

[Mishra08] Mishra S., and Palanivelu K. "The effect of curcumin (turmeric) on Alzheimer's disease: An overview," Annals of Indian Academy of Neurology (2008): *11*(1), pp. 13–19.

[Nabavi20] Nabavi N. "Long covid: How to define it and how to manage it," *the bmj* (2020): *370*. https://doi.org/10.1136/bmj.m3489

[Naviaux14] Naviaux R. K. "Metabolic features of the cell danger response." *Mitochondrion* (2014): *16*, pp. 7–17.

[Naviaux16] Naviaux R. K. "Metabolic features of chronic fatigue syndrome," *PNAS* (2016): *113*(37), pp. 472–480.

[Nunn2016] Nunn, A. V. W., Guy, G. W., and Bell, J. D. "The quantum mitochondrion and optimal health," *Biochemical Society Transactions* (2016): *44*(4): 1101–1110.

[ONS21] "Prevalence of ongoing symptoms following coronavirus (COVID-19) infection in the UK, April 1, 2021. Office for National Statistics, Census 2021. https://www.ons.gov.uk/peoplepopulationandcommunity/healthandsocialcare/conditionsanddiseases/bulletins/prevalenceofongoingsymptomsfollowingcoronaviruscovid19infectionintheuk/1april2021

[Pereira20] Pereira, M., Damascena, A. D., Galváo Azevedo, L. M., de Almeida Oliveira, T., and da Mota Santana, J. "Vitamin D deficiency aggravates COVID-19: Systematic review and meta-analysis," *Criticial Reviews in Food Science and Nutrition* (2022): *62*(5), pp. 1308–1316.

[Perrin07] Perrin R. N. "Lymphatic drainage of the neuraxis in chronic fatigue syndrome: A hypothetical model for the cranial rhythmic impulse," *The Journal of the American Osteopathic Association* (2007): *107*(6), pp. 218–224.

[Porges14] Porges, S. W., Bazhenova, O., Bal, E., Carlson, N., Serokin, Y., Heilman, K. J., Cook, E. H., and Lewis, G. F. "Reducing auditory hypersensitivities in autistic spectrum disorder: Preliminary findings evaluating the listening project protocol," *Frontiers in Pediatrics* (2014): *2*. https://www.frontiersin.org/articles/10.3389/fped.2014.00080/full

[Rabin14] Rabin, C. R. "A new women's issue: Statins." *New York Times*, May 5, 2014. *http://well.blogs.nytimes.com/2014/05/05/a-new-womens-issue-statins/?_r=0*

[Servan-Schreiber11] Servan-Schreiber, D. *Anticancer: A New Way of Life.* Michael Joseph, 2011.

[Schippers MC.] For the Greater Good? The Devastating Ripple Effects of the Covid-19 Crisis. Frontiers in Psychology. 2020 Sep 29;11:577740".

[Shaefer23] Shaefer, A. "Why Is Cholesterol Needed by the Body?", *Healthline*, 2023, https://www.healthline.com/health/high-cholesterol/why-is-cholesterol-needed

[Sherwin16] Sherwin E., Rea, K., Dinan, T. G. and Cryan, J. F. "A gut (microbiome) feeling about the brain," *Current Opinion in Gastroenterology* (2016): *32*(2), pp. 96–102.

[Tobin20] Tobin, M. J., Laghi, F., and Jubran, A. "Why COVID-19 silent hypoxemia is baffling to physicians, *American Journal of Respiratory Critical Care Medicine* (2020): *202*(3), pp. 356–360.

[Valencak17] Valencak, T. G., Osterrieder, A., and Schulz, T. J. "Sex matters: The effects of biological sex on adipose tissue biology and energy metabolism." *Redox Biology* (2017): *12*, pp. 806–813.

[Vasquez16] Vasquez, A. "Microbial origins of fibromyalgia." Naturopathic Doctor News & Review. April 12, 2016. http://ndnr.com/pain-medicine/microbial-origins-of-fibromyalgia/

[Whitley09] Whitley H1, Lindsey W., "Sex-based differences in drug activity", *Am. Fam. Physician*, 2009: 80(11) pp.1254-8.

[Wilson2023] Wilson, D. "Thyroid and anxiety." Wilson's Temperature Syndrome [blog] 1998–2023. http://www.wilsonssyndrome.com/thyroid-and-anxiety/

[Worby15] Worby, P. *The Scar That Won't Heal.* CreateSpace, 2015.

[Xie13] Xie, L., Kang H., Xu, Q., Chen, M. J., Liao, Y., Thiagarajan, M., O'Donnell, J., Christensen, D. J., Nicholson, C., Iliff, J. J., et al. "Sleep drives metabolite clearance from the adult brain." *Science* (2013): *18*(342), pp. 373–377.

[Zhao10] Zhao, L. "Genomics: The tale of our other genome." *Nature* (2010): *465*, pp. 479–480. http://www.nature.com/nature/journal/v465/n7300/full/465879a.html

CHAPTER 7

HEALTH THE BIGGER PICTURE

Vested Interests—The Politics of Health

There are huge, vested interests against making the sort of changes I've been referring to in this book, for example, diet, psychology, and self-healing strategies. The overprocessed foods that we consume both increase profits for the agribusiness industry and then, further down the line, the pharmaceutical executives who are only too happy to provide drugs to deal with the diseases of affluence that they cause. You would expect independent government regulation to check the worst excesses of the food industry, but sadly conflicts of interest permeate the regulatory and government bodies too, so regulation and governance is further compromised.[1]

Although this is evident in the price we are paying via poor health, few people are aware that we *do* have the ability to take back power by taking responsibility for what we can change, such as our diets and treating our bodies with respect and kindness. We can lobby for the rest, for example, ensuring good nutrition in public institutions (particularly for our pregnant mothers and children), so that we do not spread the imbalance to future generations—it is so much easier to prevent rather than cure.

[1.] This is not so much an overt conspiracy, as it is "experts" being deliberately co-opted by big business who fund their research and provide them with lucrative contracts. It is a marvelous way of making sure the model stays.

The best way we can effect real systemic change in our bodies is managing a healthy microbiome. This microbiome approach is *real* evidence-based medicine (EBM)[2] and will ensure the future of our species. We must recognize the interconnectedness of the human race and the earth. The microbiome, then, is the missing link to health and healing. Peace with ourselves and each other is the message that it sends when we reconnect with our microbial evolutionary origins. This includes the astonishing interconnectedness of fungi, plants and animals. Fungi allowed life on earth to develop and are important balancers of the microbial community both within and without [Zhao21], [Stamets23].

By cultivating this new relationship with your body, we enact a new blueprint. We come to the understanding that our body wants to serve us, not break us. According to nutritionist Alisa Vitti, it enables us to "tap into our inner knowledge." Ultimately, we break the obsession modern culture has with the "quick fix." Instead, the long path becomes our journey, as it ultimately works better and is longer lasting. This is the basis of the integrative approach which I have described in this book. So, it might be appropriate to now ask the question how did we get here? For the answer, we have to look to our recent history of science and discovery.

Beyond Germ Theory—The Terrain is Everything

As we discussed in Chapter 1, before the advent of modern medicine, it was believed that illness came about through bad air or "humors." It was with the advent of optical microscopes that we first saw bacteria and realized that they were responsible for most of the diseases we saw. This was coined "germ theory,", as promoted by Louis Pasteur and others. It has been a mixed blessing. While it enabled us to conquer many of the diseases that killed us suddenly and prematurely in Victorian times,[3] we still associate bacteria with germs and bugs, rather than an essential part of life.

Modern medicine still treats illness according to this *infectious disease model*, whereby we are *at war* with these outside elements that threaten our wellbeing. However, a truly effective medicine would be better focusing on promoting homeostasis (balance) not war—improving our inner garden the

---

[2] Much touted by the pharma-based medical approach as being what they espouse with randomized controlled trials, but it is a very limited interpretation of evidence where they only publish positive studies and suppress criticism.

[3] This was due in large part to high population densities and insanitary conditions in the towns and cities.

so-called "terrain" of our gut and microbiome so that we can provide ideal conditions for a naturally healthy balance which regulates itself. Pasteur even changed his mind on his deathbed saying, "the terrain is everything." We would do well to remember this, but modern medicine has abandoned its holistic (the true meaning of which is whole) origins as "quackery" and is shackled to the pharmaceutical model.

Impurities and toxins are ever present in our environment through food, the air that we breathe, on our skin, and even in our nervous system. Getting rid of these toxins is normally done through the bowel via feces, the kidneys via urine, lymph via the lymphatics, and the skin via sweat, but the process needs to be a balance of toxins in and out; this has become more and more imbalanced with modern lifestyles. A medicine that encouraged support for the body's natural systems of elimination and detoxification would be more beneficial and less expensive than mopping up after diseases have taken hold—and this is even more true for the lifestyle diseases discussed in Chapter 6.

Mostly, the ever-growing chronic diseases we are now seeing are a result of a *tipping of the balance out of homeostasis* through an overload of impurities and toxins, as our systems are overwhelmed. Ideally, bacteria should be helping us with this task as they are always present in these systems. Viruses also naturally live as part of the human body. We need to ask the pertinent question: what are they doing there if all they do is cause disease? Clearly, the human body has evolved with them for a reason, and it can't be so they can just infect us. We need to look at what is causing them to overgrow or become imbalanced.

Bacteria are everywhere in the environment; we live in a bacterial soup so we can't avoid them. They help us with absorbing minerals and disposing of waste and they even interact with the immune system, helping to prime it early in life. In other words, they keep us well. They are not our enemies, and we need to adjust our thinking accordingly. When we consider that our notion of "me" is, in fact an "us," then we can get comfortable with being part of one big internal ecosystem.

In summary, this new information changes how we view health and disease. They are not absolute states; they are a gradation, a scale along which we lie, sometimes toward one end or another. With any move toward disease, we need to address the imbalance first in order to heal. Sometimes that involves getting worse before we get better. For example, even a fever is important to experience (as long as it is not too high). Acute illness needs the fever to deal with an elevated bacterial/toxic stress—the high temperature helps to kill the pathogens. With chronic illness, the equivalent will be detox reactions that you may experience as you cleanse the body. For some chronic pain syndromes, there will be a period of intensity following a stressful experience—this is termed tension myoneural syndrome (TMS). The pain can even move around the body as the brain that controls it tries to keep your attention[4]—the so-called psychosomatic *symptom imperative*.

In addition, we now know that bacteria are not necessarily invading us from outside, but infections can result from an overgrowth of existing flora, for example, in ear infections for instance. That is why simply clearing out everything with antibiotics is so damaging long-term. We simply embed the imbalance further by killing off the benign flora that keeps the more pathogenic species in check. A better way of dealing with such problems is to clean up the diet (and for ear infections in particular you can use a salt spray in the back of the throat that links to the ear via the eustachian tube). Thus, we clean out the tissues that are clogged and struggling. Mucus membranes are often involved in enhancing removal of impurities/wastes. An example would be in sweating out impurities under our arms—body odor may be unpleasant, but it is not considered an infection. It is a normal part of keeping ourselves healthy, something that much modern medicine ignores as it increasingly pathologizes the body's detoxification and adaptation mechanisms.

Past Experience Changes Immunity

Whether we are healthy or not is greatly influenced by our early childhood experience and how our immune systems mature in the face of threat. Much of this priming of the immune system comes about in the first eighteen months of our lives. Children are exposed to bacteria that they will live with for their entire life. So much of later health is determined by this early growth in the intestinal flora. It also has a dedicated medical specialism—developmental origins of health and disease (DOHAD). Its basic premise is that a poor diet or early stress changes the balance of the microbiome and "also impairs the child's immune system, changing its ability to control the gut microbiome and opening the door to harmful infections that disrupt the communities further" [Yong16, p. 167]. The immune system doesn't just attack outsiders, it "stabilizes our relationships

---

[4.] Tension myoneural syndrome (TMS) or mind–body disorder was described more fully in my previous book *The Scar That Won't Heal*. It is a common feature of fibromyalgia, and other chronic pain syndromes.

with our microbes" much like a thermostat manages temperature in the home. Ed Yong, author of the fascinating book *I Contain Multitudes* has indeed called it an "immunostat" that allows you to see the dynamic nature of control. He says "over the last half-century, we have gradually pushed our immunostats to higher settings through a combination of sanitation, antibiotics and modern diets. We've ended up with immune systems that go berserk at harmless things like dust, molecules in our food, our resident microbes and even our own cells" [p. 167]. When this happens, changes then become permanent as the disruption goes past a critical point and dysbiosis becomes harder to rectify.

Vaccination

I looked at vaccines as an environmental modifier earlier, in the context of the complex politics of production and administration. Here, I want to look at the effect on the microbiome specifically. Vaccination is undertaken by injection into muscular tissue which then goes directly into the bloodstream. Unlike when we catch germs naturally via the mouth and nose (ingestion and inhalation), the viral/bacterial load with added adjuvants (deliberate toxins) in vaccines go directly into our bloodstream where they may trigger a whole host of changes in the microbiome of our gut.

In traditional vaccination, we stimulate the acquired immune system (Th2) to attack the protein components of the vaccine (artificial *antigen*) with proteins that identify and lock on to them called *antibodies*.[5] We also mount a fever and attempt to detoxify the other components. Usually, the result is that we destroy the specific variants present in the vaccine and prepare the body to recognize that in the future. All well and good— except most vaccines only have certain variants of the bacterium/virus (and indeed some are just fragments or genetically modified versions of natural viruses (as in the HPV vaccine). Where the other variants (*serotypes*) are not included in the vaccine, they are free to proliferate if they are present and this inevitably increases the prevalence and, unfortunately, also the virulence (strength) of the remaining strains. *In other words, it makes the other strains more powerful.* This is known as *pathogenic priming.*

[5.] The new mRNA type vaccines as developed for COVID-19 are a completely different technology where nanoparticles enter your cells and cause them to begin producing large numbers of antigens. The long-term effects of altering the molecular machinery in this way are unknown, but there appears to be some lodging in other tissues like the heart, particularly in younger people whose immune systems are more vigorous.

As some vaccines even include antibiotics too, the balance of the microbiome is irrevocably altered leading to overgrowth of pathogenic bacteria which, as we know, should be in ecological balance. This disastrous alteration, and the prevention of natural immunity that arises from fighting a natural infection, is leading to more chronic disease in later life. Vaccination has a clear negative effect longer-term, even if it may save lives in the short term. We need a balanced outlook here.

Children, it seems, are bearing the brunt of our ignorance and the statistics are frightening. In the United States, for instance, there are now:

- One in six with a learning disability.

- One in nine with asthma.

- One in ten with ADHD.

- One in fifty with autism (and this is projected to rise to one in two by 2030!).

- One in four hundred with type 2 diabetes (which used to be considered an adult disease).

The awareness of this alteration in the microbiome is missing in vaccine science. If you change the serotypes (surface recognition) of micro-organisms, you are altering the way in which they relate to each other, which then causes mutations of bacteria and viruses, selecting out the more pathogenic. Normally, the microbiome is in balance with its environment; it is supposed to mount an immune reaction to anything foreign via the adaptive (Th2) immune system but with artificial vaccines, it begins to get confused. The natural cell-mediated immune system (innate-Th1) is missed out altogether in the discussion of vaccination—but of course it is the first line defense system, and if the Th2 system is over-triggered by vaccines in this way the result is the body begins to misidentify its own tissues as foreign and autoimmune disease (AID) results.

We have seen this in the overreaction to infection with SARS-Cov-2[6] and the vaccines do nothing to improve this; only unfortunately set us up for more extreme reactions when we encounter the native coronavirus further down the line (pathogenic priming). This is being seen now with more rare thrombosis (blood clotting) events, particularly in the young and healthy (whose immune systems mount a strong response). You may argue that

---

[6.] ARDS—Acute respiratory distress syndrome is seen in people who are already highly inflamed.

the risks of the disease outweigh the risks of the vaccine (extreme adverse events are fortunately rare) but death does occur,[7] they are not benign but powerful interventions, over which we have very little control. And longer term we have no idea what the outcomes will be as the new mRNA vaccines are a completely new (and some would say untested) science.

We should be as cautious with vaccination as we are now with antibiotics.[8] It is now recognized that certain bacteria are mutating to become "superbugs" (such as MRSA), and we are changing our policy and practice to limit antibiotic use (although it is still used routinely in animal husbandry). This understanding doesn't seem to have quite permeated through in the arena of vaccinations, but the evidence is irrefutable. For instance, we are now seeing a lot of pertussis (whooping cough) bacteria due to these mutations from the overuse of the DTaP vaccine (a trivalent vaccine of diphtheria, tetanus, and pertussis). Vaccination is always a difficult decision to make for your child, as you are under pressure to follow the "standard protocol" from your doctor who gets incentives (payments) to vaccinate. Also, the research showing problems is often quickly refuted and therefore it is not easy to question the current approach. Nevertheless, there are expert voices out there who question the wisdom of high doses in quick succession and/or *multivalent* (usually three in one) vaccines like the DTaP.[9]

The new generation vaccines that have been developed for COVID-19 (but will be rolled out for other diseases as they are extremely lucrative), are a completely new approach that alters our molecular machinery directly. Vaccines developed by Pfizer and Moderna use fragments of RNA to go into our cells and produce proteins like those found on the protein spikes of the SARS-CoV2 virus. The body then is supposed to recognize that we have had an infection and produce antibodies (via the Th2 arm). This is simplistic and it implies two problems: if the spike proteins are being produced without the rest of the viral RNA, could there be an overreaction of the body when a native virus is encountered (pathogenic priming)? Indeed, it seems this is the case with people who have been vaccinated

---

[7.] There have been seven so far in the UK, mostly women aged 35–69.

[8.] Antibiotics are sometimes given necessarily to kill off undesirable bacteria and can make you considerably better (not always worse). However, they are overused indiscriminately in medicine and in animal husbandry.

[9.] They are mostly silenced, however, and a Google search will only reveal the "safe" doctrine. I recommend Dr. John Campbell for a truly evidence-based view: https://www.youtube.com/@Campbellteaching/.

actually getting more Covid infections than the unvaccinated (recent Australian data confirmed this).

More worrying, perhaps, is that retroviral mRNA has long been known to backward integrate into the human genome[10] (that is how we evolved), so what are the long-term implications for our human DNA when we introduce an artificial segment directly into our cells? Of course, the manufacturers are reassuring us that this viral DNA integration is impossible but how do we know—we simply haven't trialed these gene therapy vaccines for long enough to be sure.

The questions we need to ask are: are we creating super-organisms from our current practice? And are we setting children up for more chronic disease later on? To refuse to debate these important questions is a disrespectful response to their importance. The basic premise of vaccination is that it gives you an exposure to the organism *before* you get exposed from someone who is sick. But this is not really correct. You are always exposed, constantly. You cannot prevent colonization of the airways through vaccination because bacteria and viruses are present in the air[11][Simmonds17]. The premise is an incomplete one based on the idea that you only get sick when you are exposed to someone who is sick. However, you can be exposed without being near someone (you may harbor the germ already or it may be airborne), and you can be exposed but not get sick (i.e., be asymptomatic)—this is true of most germs as it is an interaction with the particular immune system of an individual that varies (the terrain).

It is now true to say that many people are now becoming silent carriers of these germs, particularly pertussis. We are told that the few "epidemics" we have had of measles, flu, and now COVID-19 for instance, are due to unvaccinated populations, but this may not be true either. Some experts have argued that most of the sufferers are in fact vaccinated or have been exposed, but the vaccine itself can *cause mutation* of the pathogen to new strains (variants). I realize this is contradictory to what you will read in all the mainstream press, from government watchdogs, or hear from industry spokespeople. They claim just the opposite, and state that anyone who argues against it is just putting other people's health at risk. This is a powerful (emotional) argument for compliance, but it is backwards, and it misunderstands the way the body and the microbiome work.

[10.] Via an enzyme called reverse transcriptase. SARS-Cov-2 is a retrovirus.

[11.] A nonillion 1031 in our air, sea, and soil is the current estimate, although debated.

It is *imbalance in the body* that is the problem. If you change the conditions within the body that bacteria are growing in, then it changes their growth. In the past, infectious diseases were common because dirty water, unsanitary conditions, and contaminated food was common. A lot of infection was spread this way as well as from wounds, starvation, and so on. Sanitation helped a lot in controlling morbidity and mortality, and it has been claimed that this is what in fact solved the typhoid problem eventually. But fear produces compliance—and the threat is very real—to anyone who dares argue. Parents are persuaded via public health campaigns and shamed if they withdraw, and doctors encouraged by widespread "incentives" (i.e., extra money from the drug companies when they complete a full protocol of vaccinations in their practice) and a fear of being struck off the medical register (lose their license to practice) if they don't comply. Researchers fear loss of funding and jobs too if they dare to publish against the mainstream. Pandemics arise during a climate of media-stoked fear, so I do question how much of the pathology is due to widespread panic. We know from the study of *psychoneuroimmunology* (PNI) that stress and the psychology of fear is very prominent in depressing the immune system.

Take, for example, the exposure to measles which, when I was young, most children survived by being exposed to and then getting through it. Although it does have some risks, particularly if the child is immune compromised, it is hardly ever fatal, and although it may reset the immune system to make it more likely to get other infections [Petrova19], it is not at all certain that that is as detrimental as has been trumpeted in the mass media [Grady19].[12] The immune suppression they conclude is occurring because of a decrease in antibodies to previous other infections may be due to the body changing its Th1/Th2 balance (toward innate immunity). Children who have been exposed to the virus (as opposed to the vaccine) may also grow to have less risk of cancers as adolescents and adults for instance. Researchers have noticed an age-related trend toward less cancer protection in the young (more highly vaccinated) compared to older age groups. Even a provaccination paper concluded that the reason younger people show lower protection toward blood cancer could be due to "the immunity that develops due to a real infection could be stronger than the

---

[12.] There are many problems with the two studies that were the base of this claim, in my opinion. First, they are assuming B-cell counts a proxy for the entire immune system response. They ignore the innate immune system. Secondly the control group against which they measure the "immune suppression" had been vaccinated, which as we know hyperstimulates the Th2 acquired immune system. The counts were also measured in ferrets not children.

immunity generated by the vaccine" [Kwon21]. And measles virus has even been used to fight cancer in oncology treatment [Russell09], [Engeland21]. It is a very confusing story and difficult to form an objective opinion due to the overtaking of mainstream media by provaccine commentators and writers.

Vaccines, although they have been an important part of conquering some infectious diseases in third world countries, have a very poor risk to benefit ratio. Some, like the MMR (measles, mumps, and rubella), have been acknowledged by the drug companies on their vaccine information sheet "risk encephalopathy (brain swelling), atopy (allergy), Sudden Infant Death Syndrome (SIDS), cellulitis, hypotonia (weakness of muscles), neuropathy (degeneration of nerves), somnolence (sleepiness), and apnea.[13] If you think this is propaganda, then check the list of side effects for yourself on the information sheet—although you won't be offered this you'll have to ask to see it or check it online.[14]

The problem seems to be that the risk of side effects is greater than that of the benefits if you consider a *lifetime health trajectory*. It is undoubtedly true that with the original single dose measles vaccine rates of other infectious diseases went down. But whether that is causal or just an association in not proven. Nowadays vaccines are often *multivalent*, that is, they are administered as combination in one vaccine. Nobody has dared question the intensity or multiplicity of the vaccine program, with the child now inoculated at one day old for HepA in the United States and the with multiple injections of vaccine continuing throughout childhood without the vaccines ever having been tested *in combination*.[15] I leave you to make your own opinion about this very divisive subject.

As this is a book about the microbiome, I will not go any further into this hornet's nest,[16] but restrict myself to discussion of how multiple vaccine dosing affects the gut. Some of the adjuvants deliberately open up the gut lining layer (i.e., promote leaky gut) in order to stimulate the immune system, but this is an uncontrolled response depending on the child's microbiome, immune status, and detoxification ability, all of which are biologically variable, depending on the mother's health and the genetic

---

[13.] This is taken from the contraindications list on the packaging insert.

[14.] You can find it at the EMC medicine compendium http://www.medicines.org.uk/emc/browse-documents.

[15.] I know you'll find this incredible, but it's true unfortunately of most drugs too.

[16.] The autism/vaccine debate raged for a few years but was vigorously debunked and has been ridiculed ever since.

profile of that child. We simply do not know the long-term effects of such an alteration. It is time to give you an alternative approach.

Naturopathic Medicine Approach

In *naturopathic* (natural) medicine, much more importance is given to encouraging a naturally strong immunity via supporting the child with good nutrition,[17] exercise, emotional support, and stress regulation. When you look at the body as an ecological system, symptoms are seen as a sign that the body is trying to restore health; a fever, for instance, in a necessary part of overcoming an infection and should not be prevented. Some serious discussion of this issue is needed, without resorting to the tactics of shame and blame. You will see in the press the current tirade against the "antivaccine movement" which will, according to mainstream opinion, "subvert global health" [Hotez15]. But this ignores the differences between immunization programs in the developing world and ours. Clearly there is a very strong argument for the former. Not all vaccination is bad, it's the frequency and variability of the receiving gut that is causing problems. The fact that we are now expected to vaccinate *everyone*, at least once a year, against Covid (and flu) is a totally new departure unprecedented in the history of humankind. Moreover, it totally ignores the ecology of that person.

Environment and education are also equally important to keeping well. Nutriment can be provided in other ways besides food and supplements— light is an important carrier of information for the body, so having a healthy outdoor lifestyle is important. There have been a lot of scary stories about sunlight causing skin cancer, but a body that has sufficient antioxidants should have less of a problem with sunlight. Sunlight is not the problem (unless you deliberately burn yourself constantly). Lack of antioxidants is the issue; again, it is the *balance in the body* rather than isolating certain "harmful" external factors. Vitamin D is so vital to so many functions[18] in the body and sunlight is the best source—sunlight is healthful not harmful if you are in balance with your antioxidant status. I will now look at more environmental factors.

[17.] Avoid packaged foods a lot of food marketed to children is of very poor quality. This is particularly important for babies for whom pureed veggies are better than carbs and fruits.

[18.] It's not even a vitamin, it's a hormone! It was misnamed initially before it was known to be produced in the skin and its full range of effects were known.

The Environment

There are so many different chemical/environmental threats out there; it is a wonder we haven't succumbed to extinction as a species already. It is due largely to the resilience of our gut flora that we have survived this far. Some would say we are heading to the fifth great extinction in the next fifty to sixty years [Attenborough20].

For instance, sewerage systems put industrial/chemical waste and sewage human waste into the same stream. In the UK privatized water companies have been doing routinely this unbeknown to the public and oversight bodies are strapped for cash and can do nothing. This is subjecting microbiome waste products to extreme selection pressures, creating a change in the natural microbiome in rivers and oceans. This instigates an ecological pressure for the bacteria to rapidly swap genes to protect themselves, with catastrophic results. a catastrophic result. Controversially, liquid waste is also sold to farmers to spray all over their crops. Did you know that farmers have higher rates of Parkinson's than the general population? [Van Maele-Fabry12]. This could be one of the factors, although most of the blame for this is reserved for exposure to pesticides like paraquat [Priyadarshi12], [Falzone00]. Apart from reducing our reliance on pesticides, we need to change the way we deal with human waste. Composting toilets could be the answer. The current solution of "out of sight, out of mind" doesn't really work anymore. It is causing accumulation of catastrophic biproducts which have very real-world effects, including extinction of salt-water fish in the oceans (some authors predicted complete extinction by 2048, although this has been debated [McKeever22]). That would be the end of us, too, as fish don't just provide us with food, but they keep the balance of the oceans, which is intricately linked to our climate and our water levels. If this is to be avoided, we need to rethink this system so that our ideas of sanitation can be made more sustainable. You might be wondering why you haven't heard this—it's not because the science isn't there, but rather because the idea is so paralyzing governments would prefer to ignore it.[19]

We need to start cleaning it up, as it is beginning to affect our health. Toxins are everywhere and we still don't know the cumulative effect of low doses in combination. In a recent American study, newborn babies were tested for the presence of five hundred industrial pollutants. They found three hundred including PCB, DDT, and even dioxins which have been

---

[19] This was predicted as early as 1991 when the results of overfishing on fish stocks first became identified.

banned for thirty years! The reason they're still there, even though we no longer use them, is they don't break down. Some are particularly bad as they are fat soluble and endocrine (hormone) disruptors. In the endocrine system, low doses can have a huge effect. So, toxicity begins even before you are born, via the placental blood supply in the womb.

Here for your information are the worst culprits with their effects:

- BPA bisphenyl-A can cause early onset puberty, linked to hypothyroidism (a silent epidemic).

- Perchlorate (a rocket fuel component) contaminates drinking water and is very toxic to the thyroid as it competes with iodine uptake. Hypothyroidism is not due to pure lack of thyroid hormone necessarily; you just appear deficient when your receptors are blocked. This is similar to diabetes type 2 and insulin.

- Dioxins in dairy products, coffee products, and so on, permanently affect women and men especially regarding lowering sperm counts. They are long lived, powerful carcinogens and build up in the food chain. It's really hard to avoid them altogether, but limit meat and milk/dairy products because they collect them as they are fatty.

- Atrazine (a pesticide) is pervasive in drinking water and linked to prostate and breast cancer. Drink filtered water and eat organic food.

- Triclosan is an antimicrobial and contaminates 90% of the water supply. It also disrupts thyroid,[20] and is in most antibacterial products. It should be avoided.

Actions

- Filter your water.[21] There are three hundred contaminants that have been identified and are toxic. The legal standard amount does not necessarily match the public health goal of safety—it is more about what is affordable from a water company perspective. Most filters are good for removing lead, but a reverse osmosis system is the best. Bottled water is not the answer—it can be highly contaminated as plastic bottles leach BPA in the sun.

[20.] A TRH stimulation test is more accurate than TSH to pick up hypothyroidism, but you are unlikely to be offered this by your GP.
[21.] If you use any form of charcoal filter you will need to then supplement with minerals as it takes out the good with the bad.

- Buy organic because it reduces your risk of pesticides, GMOs, added antibiotics, and hormones.

- Choose better products at home; beauty is more than skin deep. Review your personal care products. The average person puts about 126 different ingredients on their skin. Particularly bad are artificial fragrances (even the expensive ones) that contain phthalates. Also look at cleaning products and switch to natural alternatives when possible.

- Check the Environmental Working Group's "Dirty Dozen" guide. Read the labels of what you are buying.[22]

Greed (the profit motive) is at the heart of all this damage.[23] Companies are not required to disclose the damage their products cause. And it's hard to prove who is responsible, as by the time it shows up in your body it's hard to trace the origin. Testing your body burden is expensive and some of the tests are difficult to do and not particularly standardized; there are some reasonable hair tests for lead and mercury, or you can do provocation tests (where you are 'provoked to release toxins) in urine. But there is no sure way of knowing how you got certain toxins and linking them to symptoms. It is the cumulative burden that is the most critical factor and that depends on exposure and our ability to detoxify which varies from person to person.

The law, as it currently stands, is woefully inadequate. It is best to not assume anything is safe just because it is on the market. This is generally assumed by people but not true. Nobody (least of all the companies producing them), is studying the combination and synergy of these products. They are a ticking time-bomb.

Equally, when your microbiome is not healthy, you have less resilience to the toxins that are present. The antibiotics in meat are a huge issue. Food-borne illness is common and that is largely down to how the animals are raised. In the United States, seven out of ten animals are put on antibiotics for animal fattening and to make up for overcrowded conditions, which ensue from the unnatural conditions in which they are reared. It is not so bad in the UK where animal welfare protection is a little stronger (thanks to

---

[22.] See EWG website. https://www.ewg.org/foodnews/dirty-dozen.php

[23.] It is believed that support of pharmaceutical research provides a country with competitive advantage on vaccine development. During the Covid pandemic four companies battled it out for supremacy: Astra-Zeneca (UK), Moderna, Pfizer, and Johnson & Johnson (US). But these make huge profits for the corporations and not the population generally (a capitalist model which seeks only to concentrate wealth in the hands of a few).

the EU), but still no one considers the stress of the animals you are eating—their stress hormones become yours when you eat them.

To summarize, health and the environment are intimately connected, and we need to shift our inner and outer ecology toward one of harmony. This involves rejecting goods and services that are damaging to us or the environment and using more sustainable practices, which are usually more expensive. This is directly anti-ethical to the profit motive of major corporations, for whom short-term gain is the most important. However, such is the power of the consumer that it seems there is now a shift to 'slow food' and local production on more ethical lines. That shift is happening.

Furthermore, we need to reduce our fear of microorganisms—this is based on our past approach to medicine whereby we suffered huge losses from infectious plagues. This is a collective trauma, according to Martha Herbert MD, that still informs modern medicine. We need to find other ways to deal with infection and reconceptualize it as a severe acute ecological imbalance. We need to look at adding more probiotics when someone is sick, adding herbal antibiotics to rebalance the system, rather than attacking one particular process or bacterium. We are sorely going to need this within a few years, as antimicrobial resistance begins to bite. The signs are already there. We only have eight antibiotics that work broad spectrum. And that will decrease year upon year as there are no new antibiotics in the pipeline. This has been likened in the popular press to the worst pandemic we're ever likely to face and "the World Bank has estimated it could cost the world economy $1 trillion every year after 2030" [Davies18]. Government response in the UK resulted in the set-up of the Anti-Microbial Resistance Centre in 2016.[24] We need to encourage the restoration and recovery of the intestinal and oral microbiome *by all means necessary*. I leave you to ponder this while I now turn to more esoteric matters.

Mind, Matter, and Consciousness

Here I want to consider some new concepts in health science which are not usually covered in consideration of health and are overturning the old view:

Table 7.1. New Concepts in Health Science

Old Understanding	New Understanding
The body is a machine, it works perfectly until bits break down, that is, if you don't have a disease then you are healthy.	Health is a continuum, and the body is a complex system that operates as a whole not parts; as a system not organ-based model.

(Contd.)

[24.] Anti-Microbial Resistance Centre. https://www.amrcentre.com

Old Understanding	New Understanding
Disease is an unlucky event, limited by genes.	Disease is a manifestation of a dysfunctional system of which genetic predisposition is a part not the whole.
Mind is the purely the function of biological activity of the brain. It has no external validity and is limited to the brain it is housed in.	The mind seems to be a nonlocal (not limited to space–time) concept, a hologram of universal consciousness, independent of the brain. This is now established science in the world of physics [Laszlo04], but not yet in medicine.
You are born with all the neurons (nerve cells) you will ever have, and they gradually decay throughout life for unknown reasons.	Brain neuroplasticity (growth) is now known to occur throughout life; we adapt our neuronal function depending on environment.
We are only shaped by the effects of our own life experiences, a random uncontrollable combination of fate and circumstance.	Our ancestors' experiences also shape who we are via the epigenetic transmission of DNA variation—largely through the microbiome, and this may affect our life choices.

Let's look in more detail at that final, very important point. The language of DNA is the same in any creature; we are just different expressions. You could say we are "variations on a theme." The four "letters" (bases) in the DNA code make the alphabet of all life forms; dinosaur, banana, human. Genes are the words and variation depend on how we string the words together; 65% of human genes are the same as a banana, 95% the same as a fly. To paraphrase Deepak Chopra, we are just another story in the lexicon of life. But, as he and others have discovered, we can cowrite that story as we *are* consciousness itself. So, we can change the way our DNA is expressed— through epigenetic changes, largely mediated through the extraordinary contribution of our microbiome, but also through mitochondrial DNA (mtDNA). So, this can happen in our lifetime; humans have the ability to "jump" states from illness to wellness, from fear to joy—but it takes a very powerful reconfiguring of our mind-based belief systems—a "reboot" if you like. This is not easy, and it may take illness or near-death experience (NDE) to achieve.

This requires a new definition of mind, far from being related only to our brains, it permeates our whole being. Dan Siegel, a top neuropsychiatrist in the field of trauma, describes mind as: "an embodied and relational process that regulates energy and information" [Siegel23]. This is a much more powerful concept than simply the balance of neurotransmitters in our brains. It is humbling to think we are all of us interconnected; indeed, we are using recycled material from the environment around us; so, the mind is inseparable from the atoms of the universe. We are at the beginning of this new expansive understanding of ourselves, but it is not widely accepted in the mainstream, that is, outside of quantum physics.

It is understood in the history of science, that if information does not fit the prevailing paradigm, it is either ignored, ridiculed, or the opposite promoted (a defensive strategy). For instance, look at the time it took before smoking as a cause of lung cancer was accepted—and the lengths the industry went to disprove/obfuscate and generally protect their interests. Currently it is becoming accepted that exercise and diet can improve brain function to prevent neurodegenerative disease—most people are simply unaware of this. It is astounding news for a culture that simply accepts Alzheimer's disease as an inevitable consequence of aging that we can do nothing to prevent. Or the idea now gaining ground that mental health is affected by diet—a new concept that has recently been proven on mainstream TV in the UK [Chatterjee17]. We are profoundly affected by lifestyle factors, that is, the way we live, and it is up to us to change this. This is indeed being used now as a new way of approaching medicine called "lifestyle medicine," and it is set to change our current paradigm away from getting experts to "fix you" and looking to what you can do to decrease your risk of disease and improve symptoms if you already have a disease. I would like to look now at the more mental and emotional aspects of this change in approach.

Rules of Life

Here are some other ideas to change your way of being in the world: the so-called the "rules of life":

- Show up (in other words be present in the moment, even when it is difficult emotionally).

- Tell the truth (as above).

- Be quiet (allow time for contemplation and processing).

- Don't be attached to the results (the Buddhist mantra of nonattachment, one of the most difficult to achieve, we all want to "fix it" now!)

This is a practice not an end-game—in our modern culture of achievement fixation it is very difficult to allow time for *being* as opposed to doing; many fall at the first hurdle as they find being still/quiet almost impossible. However, it is possible if you find something you love that allows you to do this: coloring, gardening, being with animals/nature, and so on. This is probably the best way to achieve results, unless you find meditation easy.[25] Whatever you do, do it regularly, the more you practice these active states of mindfulness, the more the brain changes (more alpha wave activity, neural growth in areas of the anterior cingulate cortex (ACC) for instance). We become more tuned in to ourselves.

The microbiome responds to this activity via a biofeedback mechanism of energy/molecular communication. Experiments have shown that by using their intention, people can interact with their own fungi and bacteria both in experimental conditions (e.g., in a petri dish[26]) and in their own body. We share our body with them after all—you could say they *are* our body. Bacteria are not the bad guys but are responding directly to our own inner dialogue. We are part of a bigger whole and thus can't secede from nature. Thus, our microbiome, being symbiotic in nature, becomes our greatest teacher. Sometimes that is why illness is transformative—we wish it were possible through joy, but it is often through adversity, that our greatest teaching lies [Williamson15].

We need to develop better communication with that which is greater than the self, that is, our healthy essence. Call it intentionality, prayer, or some other spiritual practice, but they all lower incidence of major diseases and allow us to live longer. This development of a spiritual practice shouldn't be a freaky new-age idea that you'll get round to if you have time. It is *supremely important for your health*. Ancient cultures, with their shamans and healers, knew this. They had a much broader perspective of connectivity with the earth and our ancestors. But we have left that behind in our complete denial of the spiritual and an overfocus on attainment and material possessions. It is to our extreme detriment as a culture and species.

However, you do it, make a commitment to doing something in your life that takes you outside of your left-brain, achievement orientation and do something *without purpose*. Yes, that's right do something just for the sake of it! I know that strikes terror into some people (myself included). Coloring is a good example, but so would be walking in nature, volunteering, or singing in a choir. Whatever it is you choose to do, make it a regular

---

[25.] Meditation, although very beneficial, is very difficult for anyone with a trauma history where being still is interpreted by the body as dangerous, that is, immobilization with fear. For these people a more movement-based activity is recommended, for example, martial arts, yoga, tai chi, among others, until such time as the body is able to reset itself.

[26.] See the experiments done by the HeartMath Institute. www.heartmath.org.

part of your life. See it not as a waste of precious time (I know this is how most of us view such things), but as a means to expand your sense of time. Remember the feeling of being a child where you would just do things because you enjoyed them—that is the feeling you want to get back to—it is an expansive state and your brain and heart know the difference as they become coherent (in sync with each other).

In getting the right mindset, we find the themes of surrender and immortality, a balance of focus and letting go. Learning to "accept your body even if you are ill" helps you to accept your fear which just accentuates your suffering and ultimately comes from the fear of death. Fear of death underpins every major fear, and it is coded into our bodies to avoid—it is not easy to come to terms with our mortality, and illness may scare us into contraction and shutdown. But my experience (both personal and with my clients), has shown me that in order to heal, this fear must be faced if it is to be overcome. According to trauma researcher Peter Levine, "the fear of experiencing terror or rage (..) keeps the human immobility response in place." [Shapiro10, p. 104]. We need to go through it to go beyond it. A belief in the afterlife, if you have one, at least gives one hope that some aspect of us is infinite, therefore some part of us survives. This is one of the major lessons that come out of the studies of nonlocal phenomena. Some call it a soul, universal consciousness, the Akashic field; but it doesn't matter what word you use. You'll have to settle for the idea that some part of you is immortal. If this sounds close to a religious philosophy, well that has been the outcome of modern consciousness studies; proving that ancient mystics probably had it right after all. At the very least, science is showing us, finally, that we are more than our bodies as machines in a mechanistic world.

Spontaneous Healing

One of the most difficult things to explain using conventional medical understanding is the existence of spontaneous remission of certain illnesses. There are many examples of verifiable cases; most doctors will know of some but don't talk about them for fear that they will be accused of misdiagnosis.[27] Spontaneous healing is rare, but examples do exist; we know that they happen, but we can't force or control them. Many studies

---

[27.] Conventionally, if someone recovers from a terminal illness, the accusation of misdiagnosis will be leveled, that is, they couldn't have had it in the first place. You can see the fear behind what we can't explain. But some examples survive that have been well documented. See the book *Dying to be Me* by Anita Moorjani, for instance.

show that they are possible, and they do happen even though disdained by conventional medicine. Hersberg and Reegan collected thousands of cases in their book *Remarkable Recovery* [Hirshberg95]; you find them in every disease type. Therefore, there is a reason for hopefulness, even in the face of illness. However, that isn't to say that there is a set formula you can use to control them as there seems to be no pattern to these remissions. However, on the plus side, they are not limited to religious people; anyone can do it. It's about transcending the small (proto) self. It is perhaps more difficult with diseases that are more extreme like cancer, so I have to be careful how I say this. I am not saying that people who were unable to cure themselves are to blame, as these states are largely *unconscious*. Healing is not necessarily the same as curing anyway. Sometimes incredible healings come about through illness and sometimes even death. There is no simple formula.

As I discussed in more detail in my last book, *The Scar That Won't Heal*, modern physics is showing us that consciousness and matter are not separate as we imagined; matter requires consciousness to observe it in order to exist. Before it is observed it is just energy encompassing all possible manifestations.[28] Matter is a physical experience in consciousness, and we can't localize consciousness—it is therefore "nonlocal" and unified (matter and energy are one). This unification is even present in your body; feelings, thoughts, and images are subtle energy forms. Mind and matter, therefore, are subtle modulations (modified forms) of primary reality. We find that they are just complementary aspects of a deeper unmanifested reality like the wave and particle; two sides of an imaginary coin if you like. The unmeasurable or "subempirical" dwells in the unknown 90% of the world, where mind and matter are inseparable from atom and galaxy if that doesn't blow your mind completely! Again, I refer you to writers such as Ervin Lazslo [Laszlo16] and Alice Bailey [Bailey12] if you wish to dig deeper. The latter is a more philosophical and arcane text as it was written over eighty years ago but is still far ahead of its time.

What we now know is that our brains can connect with other brains, and this is not limited to the same space/time matrix, in the way that our lives normally are, tells us something about the nature of the human mind. We are not limited to the functions of the brain within our body.[29]

[28.] Called "collapse of the wave function" in quantum physics.

[29.] See also Ervin Laszlo's *The Akashic Field* and Larry Dossey's new book *What Is Consciousness?* The latter is too recent to be reviewed, but likely to be brilliant based on previous works he has published.

When we become aware of this and learn to train our minds in this kind of function, we learn new skills which could include healing ourselves. Modern medicine does not recognize these ideas. Physicalist ontology (the study of the nature of being as rational, physical and material) begins in medical school. Students learn anatomy from a dead body, before they progress to looking at malfunctioning parts. They learn to think of the human body as a "thing" to be subdivided into organs and tissues with no relatedness and that health is the absence of disease in one of these parts. However, your body and the universe are, in fact, a process; "a verb not a noun" if you like. We create our health with every thought we think and the way we interact on a daily basis with our environment. The physicalist limited view of medicine is missing both the animation *and* the importance of the microbiome.

One practical ramification of this idea is that with every breath you breathe in contains1 x 1022 atoms (that's trillions to you and me), so you are renewing with each breath. A million atoms that were once in the body of other people and every other species is now in you. What an amazing thought. Deepak Chopra has gone further and explains we are the total universe recycling at this moment. If so, it is true to say that you are never the same one moment to the next. And, this happens in combination with our interaction with our microbiome; so the genome—epigenome—microbiome is a continuum. We know from the study of psycho-neuro-immunology (PNI) that mind and body are not separate. Indeed, we can now speculate that even our genes are animated; there is intelligence in them.[30] In their interaction with consciousness, they are continually exchanging information, usually through the medium of light. So, we need to know that we can change our body through changing our consciousness, as well as changing the physical reality. Indeed, with a full knowledge of the central importance of consciousness we know that there is no such thing as a separate, independent material reality; they are intimately linked. Medicine may be about to catch up, but only kicking and screaming.

Medical Implications

This viewpoint of our connection within the wider world offers us a more *hopeful* view of life. The "survival of the fittest" world view is losing ground; the survival of the whole is more important. Kindness (*and I would add including kindness to the self*) is the medium. Love, growth,

---

[30.] See Deepak Chopra's *Super Genes* book.

and transformation are the message. If we allowed this paradigm shift, medicine could resume its connection with a spiritual reality to life and medical training could reflect this, rather than the current one of making supremely skilled technicians with no interest in the *human being* with the problem. When a doctor reflects that he "doesn't have time for asking the patient how they are" [Levine10, p. 108], then we know we are losing human connection in this profession. Given that 90%of healing is about belief—the so-called "placebo effect" [Dispenza14], then we are in a losing game as far as healing is concerned.

The future of medicine, then, is open—we can see it is likely to go very high tech with innovations in surgery and drug delivery, but hopefully also there will be insights into the nature of consciousness. If I hope for anything, it is that there will be substantial ethical and moral advances and a willingness to engage with the data that is already there supporting the importance of a spiritual/emotional connection to your body and each other. This is a much more uplifting message, instead of the current obsession with survival at all odds (even with no quality of life and multiple chronic conditions[31]).

To quote the great Persian poet Rumi "consciousness sleeps in minerals and dreams in plants, wakes in some humans." To understand that you are a community of 2,000 different species, living in harmony (hopefully), changes the nature of the existential question "who am I?" A more pertinent question would be "who are we?" If we go on destroying the microbiome (so far, we have destroyed 30% through pesticide use, GMOs, etc.), we kill ourselves too—slowly. Many philosophers, scientists, and clinicians are coming to the same conclusion that this slow destruction of our internal balance is probably causing the modern epidemic of diseases.

Consciousness and Purpose in Nature

So, what defines purpose in nature? There have been many advances in our understanding of this concept within current science; specifically, complexity theory. Basically, this describes that within any system that has complexity, there will also be randomness, or unpredictability (sometimes also called chaos theory). This is a nonlinear system, open to feedback that results in changes—or *emergent properties* as a creative solution to the problems presented to it. Emergence is characterized, such that what

---

[31.] As I was writing this, there have been several high-profile cases of people with little quality of life being allowed to die—some elect this whereupon it is called euthanasia. But it is still rare—we normally prolong living as we fear death.

emerges is always of a higher order than that which it emerged from. This is particularly true of biological systems subject to biofeedback. Due to its complexity, biofeedback changes within the system helps to further its own evolution. Chaos theory should be more properly called "creativity" and it is present throughout natural systems; they are not blind, impervious machines but *dynamic informational exchanges.*

Consciousness is the underlying ground of this chaos system which makes it simultaneously sentient, unitary/complementary (mind and matter are one) and recursive (i.e., it feeds back to engender higher order), so the purpose of nature, you could say is evolution, higher order, self-awareness, and enlightenment. We can recognize that there is purpose in nature even if we do not agree who or what made it (we scientists are generally careful to not define creation as a literal interpretation of the Bible). This teleological (defined by purpose) evidence would suggest that the universe is evolutionary; constantly changing, hopefully for the better.[32] The idea that there is purpose in nature, matter, and your life, has traditionally been ignored within science which has set itself up in opposition to this notion. It prefers to see evolution as random and nihilistic, guided by "survival of the fittest," which is unfortunate as this type of science has largely replaced belief in God as a religion in our modern lives, and so leaves many without meaning.

Without getting into a long, drawn-out debate on whether God exists, I would like to just point out that this is a logical fallacy; we can't prove something that is nonrational and nonlogical. That is why it has always eluded our scientific thinking. Nevertheless, some schools of thought posit that universal consciousness/God could be seen as an expression of *us.* Quantum physics has now proved that consciousness is primary, everything else is a delusion of thinking. Our whole purpose could be said to be to further the evolution of consciousness in material form, through experiencing this life. Biologically speaking, every cell has a purpose different from other cells, but all belong to a higher order of being, a whole. It is the development of the whole through the parts (including the microbiome) that is evolving. Perhaps this is purpose enough.

We have tended to put our faith in science, because it creates technology that appears to make our lives easier and better. Science is objective so it must be the truth. It is based on a supposed rational observation of

---

[32.] See *Chance and Necessity* by Nobel Prize Winner Jacques Monod (Vintage, 1971) for more discussion of this.

something solid "out there." But how do we describe a thought or inspiration that way? We can't see it or touch it; indeed, we can barely measure it, but we all know when we experience it, so we know it exists. We take an activity and call it "thought" for example when, in effect, it is process. Our technological understanding creates the map of what we think of as objective truth but, as the study of neurolinguistic programming (NLP) will tell you, "the map is not the territory," it is merely an approximation. Science is, in fact, not objective but an activity of human consciousness, so it can be divine or diabolical; think of for example the internet or the splitting of the atom, both capable of alleviation of suffering and of huge damage. Science is only a tool for organizing our experience, even though we don't know how this experience occurs. For all our understanding of the natural world, we can't know what it is like to experience the world through an animal's consciousness for instance.[33] We humans have a consciousness limited by the beliefs formed from our life experience, and that of our parents. We are conditioned by those experiences to believe we are singular beings, and the quality of our existence is defined by our efforts alone.

This "conditioned" mind has lost memory of its wholeness. Perhaps one of the few times we gain that feeling is in the service of others. If a human being acts truly selflessly to benefit another person, then we distil an essence of being bigger than our individual minds; some would say we "become God" (for example as written in the teachings of the Kabbalah and Bible). Thus, service for the good of others becomes service for the whole—this is not the same as volunteering to fill a gap in one's own life or to feel important, by the way. To know we are one with our microbiome, our environment and our ancestors is perhaps the closest approximation we have to being with God. We have that potential when we lose our illusion of self and move toward integration of the whole.

Changing Our Microbiome Psychologically

So how can we learn to speak the language of the cell/microbiome to make changes?

There are so many ways we can do this from better quality sleep, meditation, movement, enhanced mind–body coordination (learning the language of our body through yoga, for instance), and specifically, learning

---

[33.] Or can we? Some shamanic rituals, aided by psychedelics such as ayahuasca, claim to allow us to do just that.

to enhance emotions like love, compassion, joy, nonjudgment, and peace.[34] Food choice, in particular, is a major epigenetic modifier and so nutrition should not be thought of as just about calories and energy intake. When I studied nutritional medicine, this was a major revelation to me—although I had an inkling that certain foods promoted inflammation for instance, I didn't realize quite how finely tuned the whole system of food intake, microbiome, and health is. Food alone can heal you—it's just that tackling the other factors at the same time will make it faster and more enduring.

The microbiome is also influenced by our lives, as we have seen. They form communities called guilds where they all speak to each other, but there are a few that are particular leaders called "Alexander organisms" (those organisms that can organize and direct other microbes); examples are Plantarum spp. and Lactobacillus johnsonii.[35] The microbiome is not just bacteria, it is very diverse and full of viruses, bacteriophages that control the viruses, protozoa, archaea (previously thought to be a bacterium now reclassified), and fungi; all produce endotoxins and thus inflammation, which, as we've begun to realize, is a feature of all main diseases.

Inflammation is best thought of as "a gun that's ready to go off." What pulls the trigger is our lifestyle, for example, a high-carb diet, too many of the wrong fats, toxins (heavy metals, radiation, etc.), and toxic thoughts and beliefs. Diet is the most importance epigenetic modifier of all because once it changes the balance of species in the microbiome and they, in turn, make you hungry for foods they prefer. This can get stuck as a habit, as these bacteria then release neurotransmitters that keep you craving these foods. So, in any change program, it takes at least two days before the microbiome message of cravings is reduced. Getting through the cravings is the most difficult part, but once you have done it, you can begin to alter the balance by eating good foods.

Using the Power of Mind

The study of human consciousness has come against many unsolvable questions; what makes a human being able to be aware of itself? This is the so-called "hard problem" that has largely been ignored by modern psychology, and that has concerned itself primarily with cognitive processes and behaviorism. However, with recent advances in scanning

---

[34.] In my years of experience working with people, they mostly send the opposite messages of distrust, threat, and judgment toward their own body. We have not learned to love and support ourselves well.

[35.] See the work of Leo Galland.

technology, we have been able to map the neural correlates of certain functions of core consciousness within the brain, for example, sensory perception, attention, memory, and so on. This model of the mind which sees it as being housed only in the brain, as a function of its activity (neurons exchanging information mechanistically), does not explain certain "nonlocal" (i.e., not limited to the brain) manifestations of consciousness, for example, precognition, extra sensory perception (ESP), or remote healing. These abilities of the human mind to communicate outside of the body have been known about for years but ignored in the mainstream as they question the whole dogma; it has been called "one of the best kept secrets in medicine" according to influential physician Larry Dossey [Dossey02].

In order to heal, we need to make a place in our life for a contemplative practice, whether meditation or less formally with mindful awareness, walking, or creativity. This helps us get out of the mindset of aloneness and into a sense of connection with the web of life that exists within and around us; what has been termed "One Mind" by Dossey and others. Other ancient cultures had this activity in their lives routinely; in the Western world we have lost it almost entirely. But thankfully, there are a number of areas that are being investigated within medicine; one is the placebo effect (whereby a sham drug still has a positive effect), which has been thoroughly researched, not as an unfortunate adjunct of clinical trials to be eliminated, but as a positive force to be harnessed. Dr. Joe Dispenza, Ted Kaptchuk [Feinberg13], and other more forward-thinking doctors have embraced this regardless of whether the patient is aware or not.[36]

It seems that the mind, when harnessed appropriately, can enhance healing, slow aging, and promote well-being [Marchant11]. As Deepak Chopra has said, "nothing holds more power over the body than the mind" [Chopra13, para. 3]. We would do well to incorporate this into our medical model. Doctors can do this just by believing in their patient's ability to heal—the modern practice of giving patient's the worst-case scenario, (even in the form of patient information leaflets), projects in the mind of the patient the worst outcome not the best. When a doctor says to a patient, as was said to me personally a few years ago when I had a small skin cancer, "You will undoubtedly have more," this signals to the mind–body an absolute truth, which then becomes so unless challenged. Luckily, I was able to overturn

---

[36.] There are in fact studies which show knowingly taking a placebo can still induce positive changes in health. There are certain rules: certain colors of pill work better than others, and the bigger the pill the better! Strange, but true.

this mind-based program with a stronger belief in my body's ability to heal given the right information. The mind–body, it turns out, is not solid, but a process in motion, one that we remake constantly. That is why it is important to give it the right messages and having connection with a deeper meaning can enhance that many-fold. That is why people with a strong spiritual belief often live longer despite illness; even with diseases like cancer their survival is enhanced. Don't forget the messages we send are picked up by *all* our body's cells (including those of the microbiome).

Spirituality

You may be surprised to have a section on a subject such as spirituality in a book about the microbiome. However, any gut problem has components of emotional issues tied to our daily lives. For instance, anxiety often manifests physically in the gut—it churns or becomes stagnant and constipated. The gut and brain are connected in ways that we are only now beginning to understand. If we stress out over lack of control over life events and feel a failure, this is a spiritual problem. It is often driven by fear—our fear-driven culture inculcates a deep fear of the future, fear of death (particularly of cancer of Alzheimer's where we lack control). This has powerful effects on our microbiome if we live by that fearful thinking.

The antidote to this kind of life is allowing; letting life happen and embracing your situation, even if it involves suffering without having to "fix it." I am aware this is a great challenge to most people, but having a spiritual connection to the bigger picture allows us to get a different perspective on what is going on for us right now, even if it involves suffering, is the key to recovery. Surrender to what is becomes a journey in self-discovery, even to finding out what is good in the situation, learning from embracing it until eventually you find that you have grown immeasurably from the process. After a while you get to the point where you wouldn't trade it because it has given you such inner knowledge. This is where I have come to in life. Not always, but to some extent, I have found meaning in even the not so enjoyable moments. I have grown spiritually in other words, even without being religious or devotional in any sense. We need to take back the idea of spirituality as a connection not as a religious meaning, although that is fine too if it works for you.

Love and Gratitude

Love and gratitude are the most powerful healing emotions. Your purpose is to find joy in your life. Sadly, when we get pressured or ill, we often drop those things that we love the most. Don't squeeze those things out; embrace all the things and people that give you meaning. Remember whose life you have impacted for the better.[37] This awareness helps to heal the gut and the body. So "love your fate!" Live to the fullest integrity of who you are and be authentic—even if it upsets some people. So many people end up not being themselves but what someone else thinks they should be (usually someone whose approval/love they needed in childhood). They live their lives like this, and their body tells them in no uncertain terms that this is inauthentic. Often, this manifests in pain or fatigue, a feeling of malaise. Feeling love toward oneself as you would a child helps to counterbalance this. If you have an active religious spiritual practice, then the power of prayer may also be invoked.

Health is completely multifactorial and so isn't just about the physical aspects of living. We mustn't forget the role of love and emotional connection too. Science has begun to embrace this idea as a means of reducing stress in the population and has called it "positive psychology." Its basic premise is the need to increase gratitude, holism, empowerment, and ultimately, love by practicing these behaviors consciously. With regular practice, you get more desire and fierceness for *life* (not the artificial one for money and status). You can't get this by taking antidepressants; although they may help you get through a difficult time, ultimately, they take away ambition. I've seen many people in my practice who have lost the fire for life and that is sad and unnecessary. We can rewrite this story; we can learn to take care of ourselves and our microbiome to tap into their power to heal us. So, it's no longer about what "I" can do but "we." This is a very important understanding.

Toward a Twenty-First Century Healthcare System

We don't have a healthcare system at the moment; according to many experts, and, thankfully, now doctors too,[38] it is beginning to be recognized we have a disease management system. A system that actively encourages sickness and disempowers us from taking active measures to heal ourselves.

[37.] When I have something go wrong or I am criticized, I often pay much more attention to these things than all the good deeds I have done. This is a default of the brain to overaccentuate the negative. But by reminding yourself of the positive effect of your life is a valuable practice. None of us are perfect, but we would do well to accept ourselves on a fundamental level—being kind to ourselves is a spiritual practice.

[38.] See, for example, the manifesto by Dr. Sanjay Gupta, known as York Cardiology on YouTube.

What would a twenty-first-century healthcare model that incorporated all this new information look like? Here are some suggestions:

- Preventative—not a disease "fix it when its broken" model but one that looks at prolonging healthy life through lifestyle and a "systems approach."

- Participatory—you have to *do* something and take responsibility for your health.

- Personalized—no one size fits all, whether it's genetics or diet, each of us is different.

This would be a collaborative process between you and your healthcare provider. The healthcare system becomes an active, energized, empowered process rather than the sad, sick, disease model we currently have. And it uses natural means; food, exercise, sleep, sunshine, and stress release to do it. For more information on this approach, see the work of the Personalized Lifestyle Medicine Institute.[39]

References

[Attenborough20] Attenborough, D.. *Extinction*. (London, BBC, London, 2020), film.

[Bailey12] Bailey, A. *The Consciousness of the Atom*. Martino Fine Books, 2012.

[Chatterjee17] Chatterjee, R. *Doctor in the House*. (2017). Episode 2, Series 2. BBC Programs. http://www.bbc.co.uk/programmes/b08rcjdb.

[Chopra13] Chopra, D. "Harness your mind's power to heal and transform." chopra. June 11, 2013. https://chopra.com/articles/harness-your-minds-power-to-heal-and-transform

[Davies18] Davies, M., Adams, C., and Newell, C. "The true cost of antibiotic resistance in Britain and around the world." *The Telegraph*, March 29, 2018. https://www.telegraph.co.uk/global-health/science-and-disease/almost-died-true-cost-antibiotic-resistance-britain-around-world/

[Dispenza14] Dispenza, J. *You Are the Placebo: Making Your Mind Matter*. Hay House, 2014.

---

[39] PLMinstitute.org

[Dossey02] Dossey, L. Healing Beyond the Body: Medicine and the Infinite Reach of the Mind. Sphere, 2002.

[Engeland21] Engeland, C. E., and Ungerechts, G. "Measles virus as an oncolytic immunotherapy," *Cancers* (2021): *13*(3), 544.

[Falzone00] Falzone, L. et al., "Occupational exposure to carcinogens: Benzene, pesticides and fibers." *Molecular Medicine Reports* (2000): *14*(5), pp. 4467–4474.

[Feinberg13] Feinberg, C., "The placebo phenomenon," *Harvard Magazine*, January/February, 2013. https://harvardmagazine.com/2013/01/the-placebo-phenomenon

[Grady19] "Measles makes your immune system's memory forget defenses against other illnesses." New York Times, November 19, 2019. https://www.nytimes.com/2019/10/31/health/measles-vaccine-immune-system.html

[Hirshberg95] Hirshberg, C. *Remarkable Recovery: What Extraordinary Healings Can Teach Us About Getting Well and Staying Well.* Headline Book Publishing, 1995.

[Hotez15] Hotez, P. L., "Will an American-led anti-vaccine movement subvert global health?" [Blog] American Scientific, March 3, 2017. https://blogs.scientificamerican.com/guest-blog/will-an-american-led-anti-vaccine-movement-subvert-global-health/

[Kwon21] Kwon, K. "Many people with cancer lack protection against measles and mumps," Inside Science, August 12, 2021. https://www.insidescience.org/news/many-people-cancer-lack-protection-against-measles-and-mumps

[Laszlo04] Laszlo, E. Science and the Akashic Field. An Integral Theory of Everything. Inner Traditions, 2004.

[Laszlo16] Laszlo, E., and Chopra, D. *What Is Reality? The New Map of Cosmos, Consciousness, and Existence.* New Paradigm, 2016.

[Levine10] Levine, P In an Unspoken Voice, How the Body Releases Trauma and Restores Goodness. North Atlantic Books, 2010.

[Marchant11] Marchant, J. "Heal thyself," *New Scientist,* August 24, 2011. https://www.newscientist.com/article/mg21128271-900-heal-thyself-meditate/

[McKeever22] McKeever, A. "How overfishing threatens the world's oceans—and why it could end in catastrophe." *National Geographic Magazine*, February 7, 2022. https://www.nationalgeographic.com/environment/article/critical-issues-overfishing

[Petrova19] Petrova, V., Sawatsky, B., Han, A. X., Laksono, B. M., Walz, L., Parker, E., Pieper, K., Anderson, C. A., de Vries, R. D., Lanzavecchia, A, et al. "Incomplete genetic reconstitution of B cell pools contributes to prolonged immunosuppression after measles," *Science Immunology* 4(41), article 6125.

[Priyadarshi12] Priyadarshi, A., Kuder, S. A., Schaub, E. A., and Shrivastava, S. "A meta-analysis of Parkinson's disease and exposure to pesticides, *Neurotoxology* (2012): 21(4): 435–440.

[Russel09] Russell, S. J., and Peng, K. W. "Measles virus for cancer therapy," *Current Topics in Microbiology and Immunology* (2009): 330, pp. 213–241.

[Shapiro10] Shapiro, R. The Trauma Treatment Handbook: Protocols Across the Spectrum. W. W. Norton, 2010.

[Siegel23] Siegel, D. "Mindsight summary." Briefer, 2023. https://briefer.com/books/mindsight

[Simmonds17] Simmonds P., Adams, M. J., Benko, M., Breitbart, M., Brister, J. R., Carstens, E. G., Davison, A. J., Delwart, E., Gorbalenya, A. E., Harrach, B., et al. "Virus taxonomy in the age of metagenomics," *National Review of Microbiology* (2017): 15, pp. 161–168.

[Van Maele-Fabry12] Van Maele-Fabry, G., Hoet, P., Vilain, F., and Lison, D. "Occupational exposure to pesticides and Parkinson's disease: A systematic review and meta-analysis of cohort studies." *Environment International* (2012): 1(46), pp. 30–43.

[Williamson15] Williamson, M. Return to Love; *Reflections on the Principles of a Course in Miracles*. Harper Thorsons, 2015.

[Yong16] Yong, E. I Contain Multitudes: The Microbes Within Us and a Grander View of Life. Bodley Head, 2016.

[Zhao21], [Stamets23 Zhao S, Gao Q, Rong C, Wang S, Zhao Z, Liu Y, Xu J. Immunomodulatory Effects of Edible and Medicinal Mushrooms and Their Bioactive Immunoregulatory Products. J Fungi (Basel). 2020 Nov 8;6(4):269." And "Stamets P., "How Psilocybin Mushrooms Can Help Save the World", SXSW 2023 https://www.youtube.com/watch?v=qry8K7KPHIQ"

CONCLUSION

We are finally moving toward the idea of our bodies as consisting of "the integrated activities of the resident *holobiont* (another word for microbiome) consisting of bacteria, viruses, and fungi) (...) coordinated in the meta-organism in response to various environmental pressures."[1] The language of that system is energy and information, not physical matter. So, a medicine that still clings to a "billiard ball" view of atoms and molecules, is largely out of date and increasingly irrelevant. As modern medicine struggles to meet the demands of a toxic world, old systems are crumbling. But in their place lies a world of possibility. I urge you to join in, get educated, and take control.

---

[1] Gilbert, S. F. (2014). "Symbiosis as the way of eukaryotic life: the dependent co-origination of the body," *Journal of Biosciences* (2014): 39(2), 201–209.

EPILOGUE

My first book, *The Scar That Won't Heal* (kdpe 2015, revised 2019 and 2022) was a labor of love, and I never really expected many people to read it, although I hoped they would, of course! It was born of necessity from my own travails and the understanding I had reached. This second book, like the first, was an idea that popped into my head—with the title—and a very firm instruction that I should write this. I studied nutritional medicine in 2010-12 but I don't think the microbiome was covered much, apart from in the section on parasites and infections. Such is the negative bias we have toward the bugs that live in and on us.

But, writing this book I was very keen to stress the positive impact they make despite the negative fear inducing press! I think I must be connected to some supraconscious mind because, almost as soon as I had begun compiling my book, another book came out called *I Contain Multitudes* by another Brit, Ed Yong, which seemed to indicate I was in the zeitgeist somehow. This happened during the writing of the first book too. Strange. However, I was not deterred as I still felt my approach was that understands the body *as a system*, an appreciation of the role of consciousness and intention was still not fully elucidated. I hope I have brought you to that understanding, and perhaps enticed you to carry on in your research.

As I was nearing its completion, I was diagnosed hypothyroid—somewhat ironically, considering I had been researching this extensively for the book. It was not without some consternation on my part—I do look after my gut, after all. However, stress and emotions have a huge role to play in these disease expressions of the body; if I am very honest it has been hard work supporting an elderly, ailing mother emotionally while trying to complete this book—maybe that has been a little too much for my depleted system. Heal the healer as they say, so I am now on an integrated approach of both holistic interventions (based on all the good advice in this book) and can report that I am making some progress. Perhaps, humbly, I may offer that it might warm you to me more knowing I am not just the enlightened "expert" with no grounding in real-life troubles. Since Covid, everything has altered beyond recognition, but the information herein is even more needed. I wish you well on your own healing journey as a fellow health warrior.

acute respiratory distress syndrome ARDS)—the inflammatory response of the lung tissue to COVID19 when in immune dysregulation

adaptogen—in herbal medicine, a natural substance considered to help the body adapt to stress. Includes: ashwaganda, astragalus, ginseng, liquorice root, holy basil, maca, some mushrooms, and rhodiola.

adenosine triphosphate (ATP)—the molecule of energy production and a signaling molecule produced throughout the body in the mitochondria.

antigen—the component of an invader to the body (typically a virus or bacteria but can be pollen or a chemical) that triggers the immune antibody response.

antibody—the body's protein defense molecule to attach to the antigen on an invader to enable recognition as a foreign substance.

autoimmune disease (AID)—a category of diseases consisting of immune over stimulation where the body mistakenly attacks its own tissues believing them to be foreign. Many examples exist, including rheumatoid arthritis (RA) and Hashimoto's thyroiditis.

autonomic nervous system (ANS)—part of our nervous system (alongside the central nervous system) that works behind the scenes to regulate our body without our conscious involvement such as breathing, the heartbeat, and digestion. It has two branches: the sympathetic (fight and flight) and the parasympathetic (rest and digest). It is mainly efferent, that is, body to brain.

autophagy—the natural intracellular degradation mechanism by which cells (and organelles like mitochondria) renew themselves in order to recycle their materials and keep functioning efficiently.

blood brain barrier (BBB)—a highly selective semipermeable membrane barrier that separates the circulating blood from the brain extracellular fluid in the brain. Keeps the brain free of toxins.

brain-gut-microbiota axis (BGM) —the very important neuro-endocrine links between these three organs which coordinate responses to the environment. Highly implicated in health and disease.

biofilm—a colony of microorganisms in which cells adhere to each other and to a surface in order to protect themselves. They are frequently embedded within a self-produced matrix of extracellular polymeric substance (EPS). Examples are found on the skin and in the mouth.

central nervous system (CNS) —one of the two branches of nervous system with the autonomic nervous system. This consists of primarily voluntary functions and is mainly afferent, that is, brain to body through the spinal cord.

chelator—a molecule that binds minerals and is used by the body to detoxify and remove toxins from the body. Examples include chlorella (an algae) and DSM (an organic chemical).

chromophore—a chemical high in double bonds that helps harvest light and provides food with its coloring.

commensals—bacteria that live in harmony with us, sharing resources but producing valuable products which help strengthen our immunity and build resistance against pathogenic micro-organisms.

cytochrome P450 (CYP) enzyme—a very important detoxification enzyme system in the liver, the strength of which varies depending on your genetic make-up.

digestive enzymes—the complex array of digestive proteins that speed up digestion. There are different enzymes for different food groups—amylases for starches, lipases for fats, and proteases for proteins.

dysbiosis—a general term to describe the imbalance of gut flora that contributes to disease. Hippocrates, the father of medicine, said "all disease begins in the gut" and this is now corroborated.

electron transport chain (ETC) —a chain of protein complexes in the inner membrane of a mitochondrion that allow energy production by transfer of electrons, protons, and photons.

enteric nervous system (ENS)—the nerve complexes that line the gut— also called the "gut-brain." Mostly consisting of efferent (toward the brain) fibers.

epigenetics—the study of the factors above the simple reading of genetic blueprint that dictates what genes are read and in what order. Factors include diet, tissue pH, stress hormones, and so on.

fecal transplants—the artificial transfer of healthy fecal flora from one person to another to promote health. A very new idea that is gaining ground.

functional medicine—a systems-based model of medicine that looks at rebalancing the inter-relationships between hormonal, neuronal, and nutritional systems of the body. It seeks to identify and address the root causes of disease, rather than simply treat symptoms. Although not the conventional view yet, it is rapidly gaining ground, and will, no doubt, be the future of medicine.

gastric hydrochloric acid—naturally occurring in the stomach lining (secreted by Purkinje cells) to help digest protein that needs an acidic environment.

genome complexity conundrum—the surprising finding that the human genome contained only 26,000 genes—roughly the same as an earthworm! Our complexity has been subsequently found to be due to contribution of the microbiome—sometimes called the "second genome."

genotype—the particular configuration of genes that an individual has, in other words its genetic makeup.

germ theory or infectious disease model (IDM). The idea, proposed in the nineteenth century, is that all disease stems from infection with external disease-causing agents (usually bacteria and viruses) that are transmitted between individuals and thus spread disease.

glial cells—specialized caretaker cells that insulate and help neurons in the brain keep healthy.

gluten intolerance or nonceliac gluten sensitivity (NCGS) — a common condition in which gluten causes nonspecific reactions without being a true allergy. It is very common.

gut-associated lymph tissue (GALT)—a layer of lymph vessels that sit beneath the epithelial (skin) cells of the gut in areas called Peyers patches; they drain into the lymph nodes.

glycolysis—a secondary pathway for producing energy outside the mitochondria that produces much less energy. It is a back-up system but can be switched to as a response to stress.

holobiont—another word for the microbiome.

homeostasis—a tendency toward a relatively stable equilibrium between interdependent elements, especially as maintained by physiological processes within the body.

horizontal gene transfer (HGT)—the movement of genetic material between unicellular and/or multicellular organisms compared to the more widely understood "vertical transmission" (the transmission of DNA from parent to offspring). Probably so named because genetic family trees are normally mapped vertically for the parents and horizontally for the offspring.

human genome project—launched in 1990, is a project that would fully map and sequence the human genome. It was completed in 2003.

hypoxaemia—low blood oxygen in the hemoglobin carrying capacity of red blood cells. Linked with serious COVID infection.

intestinal permeability (leaky gut) —finally accepted by mainstream medicine as contributing to disease, this syndrome of opening up of the gut barrier allows proteins to enter into the bloodstream causing the immune system to react to food as if it were a pathogen and mount an immune response.

interoception—the ability to detect your own internal neurological state neurologically; a term coined by Stephen Porges.

macrophages—millions of viruses that infect the bacteria making up our microbiome. Hence, like Russian dolls, there are bugs within bugs, which increases the genetic material of our bodes hugely.

metabolomics—the study of the set of metabolites (cellular protein, lipid, and other macromolecule content) present within an organism, cell, or tissue. A chemical "signature," in other words, brought about by advances in detection methods.

microbiome—the collective name for the microbes that live in and on your body—specifically it refers to the DNA component (microbial genome) but has now come to mean the organisms themselves.

mitochondriopathy—a failure of the mitochondria to produce enough energy due to down-regulation of the system by toxicity and stress. It's at the heart of most chronic diseases, including cancer.

neuroception—the ability to feel safe neurologically; a term coined by Stephen Porges.

neuropeptides—small protein molecules that, like neurotransmitters, communicate information throughout the body.

neuroplasticity—the term given to the modification of neuronal pathways in the brain (the chain of nerve cells) as a result of changes in the environment. Now known to occur throughout life.

noncommunicable disease (NCD) —new designation for chronic types of disease now seen more prevalently than the communicable (infectious) types of disease that used to afflict mankind. We tend not to die from NCDs, but they are the leading cause of ill health worldwide, which modern medicine struggles to solve.

oxidative phosphorylation (or **oxphos**)—a series of chemical reactions in the mitochondria that use oxygen to produce energy from glucose via the electron transport chain. It is the most efficient way of producing energy for the body.

pathogenic priming—the unfortunate result of promoting more severe symptoms when exposed to a new variant of a virus because the previous one has primed the system. This can also happen as a result of repeated vaccinations.

phenotype—the set of observable characteristics of an individual resulting from the interaction of its genotype with the environment.

prebiotic—a carbohydrate that probiotic microbes preferentially eat.

psychobiotics—a group of probiotic bacteria, which, when ingested, confer mental health benefits through interactions with commensal gut bacteria.

psychoneuroimmunology—the study of the interaction between psychological processes and the nervous and immune systems of the human body.

single nucleotide polymorphism (**SNiP**) —unlike mutations that involve whole genes, this is a small change in the DNA alphabet of bases that has one base alteration (letter). They are very common and can now be mapped to give an indication of your susceptibility to certain conditions.

skin associated lymph tissue or SALT—similar to the GALT but underlying the skin.

small intestinal bacterial overgrowth (SIBO) —a common issue, although very under-diagnosed, is the presence of species of bacteria in the upper gut (small intestine) that should not be there. They cause all sorts of issues from gas and bloating to diarrhea and inflammation in the lower gut.

stress response—a specific physiological hormonal and neurological response to perturbations of the internal and external environment to bring it back into homeostasis.

Th1 or Th2 dominant immunity—reflects the type of T-helper cells that predominate in your immune response (roughly correlating to whether the innate or adaptive type immune systems). T1-type cytokines produce the proinflammatory responses responsible for killing intracellular microbes and cancer and Th2 cells are anti-inflammatory and aim to destroy pathogens that occur outside our cells (bacteria and parasites). Imbalance in either of these branches tends to result in autoimmune disease or cancer respectively.

vagus nerve—a very important parasympathetic nerve (also called "the wanderer," originating as tenth cranial nerve of brainstem), which regulates organs above (fast, ventral vagal) and below (slow, dorsal vagal) the diaphragm.

zonulin—an important protein which upregulates gut permeability. Implicated in leaky gut and ceeliac disease. A precursor of adaptive immune haptoglobins.

INDEX